RAPHAEL S. COHEN, NATHAN BEAUCHAMP-MUSTAFAGA, JOE CHERAVITCH, ALYSSA DEMUS, SCOTT W. HAROLD, JEFFREY W. HORNUNG, JENNY JUN, MICHAEL SCHWILLE, ELINA TREYGER, NATHAN VEST

COMBATING FOREIGN DISINFORMATION ON SOCIAL MEDIA

STUDY OVERVIEW AND CONCLUSIONS

Prepared for the Department of the Air Force
Approved for public release; distribution unlimited

RAND PROJECT AIR FORCE

For more information on this publication, visit **www.rand.org/t/RR4373z1**.

About RAND

The RAND Corporation is a research organization that develops solutions to public policy challenges to help make communities throughout the world safer and more secure, healthier and more prosperous. RAND is nonprofit, nonpartisan, and committed to the public interest. To learn more about RAND, visit www.rand.org.

Research Integrity

Our mission to help improve policy and decisionmaking through research and analysis is enabled through our core values of quality and objectivity and our unwavering commitment to the highest level of integrity and ethical behavior. To help ensure our research and analysis are rigorous, objective, and nonpartisan, we subject our research publications to a robust and exacting quality-assurance process; avoid both the appearance and reality of financial and other conflicts of interest through staff training, project screening, and a policy of mandatory disclosure; and pursue transparency in our research engagements through our commitment to the open publication of our research findings and recommendations, disclosure of the source of funding of published research, and policies to ensure intellectual independence. For more information, visit www.rand.org/about/principles.

RAND's publications do not necessarily reflect the opinions of its research clients and sponsors.

Published by the RAND Corporation, Santa Monica, Calif.

Library of Congress Cataloging-in-Publication Data is available for this publication.
ISBN: 978-1-9774-0718-4

Cover: J.M. Eddins Jr./U.S. Air Force; James Thew/Adobe Stock.

Preface

How are countries using social media—particularly disinformation campaigns—to influence the competitive space? How have governments, the private sector, and civil society responded to this threat? What more can be done? And what do all these conditions mean for future U.S. Air Force and joint force training and operations?[1] This report attempts to answer some of these questions as part of a broader study of disinformation campaigns on social media and the implications of those campaigns in great-power competition and conflict. The other volumes in this series are:

- Elina Treyger, Joe Cheravitch, and Raphael S. Cohen, *Russian Disinformation Efforts on Social Media*, Santa Monica, Calif.: RAND Corporation, RR-4373/2-AF, forthcoming
- Scott W. Harold, Nathan Beauchamp-Mustafaga, and Jeffrey W. Hornung, *Chinese Disinformation Efforts on Social Media*, Santa Monica, Calif.: RAND Corporation, RR-4373/3-AF, 2021
- Raphael S. Cohen, Alyssa Demus, Michael Schwille, and Nathan Vest, *U.S. Efforts to Combat Foreign Disinformation on Social Media*, Santa Monica, Calif.: RAND Corporation, 2021, Not available to the general public.

The research reported here was commissioned by the Air Force Special Operations Command and conducted within the Strategy and Doc-

[1] This report was completed before the creation of the U.S. Space Force and therefore uses the name "U.S. Air Force" to refer to both air and space capabilities.

trine Program of RAND Project AIR FORCE as part of the fiscal year 2019 project "Bringing Psychological Operations and Military Information Support Operations into the Joint Force: Counterinformation Campaigns in the Social Media Age," which was designed to assist the Air Force in evaluating the threat of foreign influence campaigns on social media and assessing possible Air Force, joint force, and U.S. government countermeasures.

This report should be of value to the national security community and interested members of the public, especially those with an interest in how global trends will affect the conduct of warfare.

RAND Project AIR FORCE

RAND Project AIR FORCE (PAF), a division of the RAND Corporation, is the Department of the Air Force's (DAF's) federally funded research and development center for studies and analyses, supporting both the United States Air Force and the United States Space Force. PAF provides the DAF with independent analyses of policy alternatives affecting the development, employment, combat readiness, and support of current and future air, space, and cyber forces. Research is conducted in four programs: Strategy and Doctrine; Force Modernization and Employment; Manpower, Personnel, and Training; and Resource Management. The research reported here was prepared under contract FA7014-16-D-1000.

Additional information about PAF is available on our website: www.rand.org/paf/

This report documents work originally shared with the DAF in September 2019. The draft report, issued on September 20, 2019, was reviewed by formal peer reviewers and DAF subject-matter experts.

Contents

Boxes and Figures

Boxes

Figures

Summary

Issue

Ever since Russian interfered in the 2016 American election, the question of how to best combat foreign disinformation attention has attracted significant public scrutiny and policy attention. This report and the others in the series, sponsored by U.S. Air Force Special Operations Command, focuses on two sets of questions. First, how are state adversaries—particularly China and Russia—using disinformation on social media to advance their interests, and what does the joint force—the U.S. Air Force, in particular[1]—need to be prepared to do in response? Second, what are the joint force, the U.S. government, civil society, and private industry doing to combat these threats? How effective are these efforts, and what more needs to be done?

Approach

Our research approach drew on a host of different primary and secondary sources; data sets; and, perhaps most importantly, more than 150 original interviews from across the U.S. government, the joint force, industry, and civil society. To understand Russian and Chinese actions, we interviewed experts from Belarus, Belgium, Japan, Philippines, Singapore, South Korea, Taiwan, Ukraine, and the United Kingdom.

---

[1] This report was completed before the creation of the U.S. Space Force and therefore uses the name "U.S. Air Force" to refer to both air and space capabilities.

Conclusions

Disinformation campaigns on social media might be more nuanced than commonly portrayed.

- Russia and Iran have used this tactic abroad more than China and North Korea have.
- State-led disinformation campaigns on social media are a relatively recent phenomenon.
- The campaigns can intimidate, divide, and discredit, but there is limited evidence that they can change strongly held beliefs.
- Smaller, locally popular social media platforms could be at higher risk of disinformation than larger, mainstream ones.
- Disinformation campaigns on social media have clearly notched operational successes, but their strategic impact is less certain.
- Disinformation campaigns on social media will likely increase over the coming decade.

At the same time, much of the response to disinformation remains ad hoc and uncoordinated.

- The U.S. government's lead for countering disinformation, the Department of State's Global Engagement Center, lacks the necessary political and institutional clout to direct a coordinated effort.
- The joint force's efforts to man, train, and equip forces for counter-disinformation remains ad hoc and service-dependent.
- Allies and partners have tried other countermeasures, mostly with ambiguous effects.
- Industry—particularly after the 2016 election interference—has made an active effort to counter disinformation, but remains motivated mostly by bottom lines.
- Civil society groups play an important, often overlooked role.

Finally, although the disinformation campaigns on social media will likely increase over the coming decade, it remains unclear who has the competitive edge in this race because disinformation techniques are evolving at the same time as these countermeasures.

Recommendations

Air Force Special Operations Command and the Air Force

- Expand information operations capabilities and focus on more than operational security.
- Weigh Commando-Solo deployments carefully, especially in adversaries' "near abroad."

Joint Force

- Know the information environment (see Figure S.1) and focus on more than just U.S. social media platforms.
- Train for disinformation, focus on key demographics, and minimize bans of smartphone and social media use.
- Increase transparency and enforce message discipline.
- Conduct a Department of Defense–wide review of structure and authorities of the information operations force.

U.S. Government at Large

- Publish a counter-disinformation strategy.
- Leverage industry but do not outsource the counter disinformation fight.
- Leverage civil society groups.
- Avoid bans of social networks.
- Balance countering disinformation with commitment to freedom of speech.
- Focus offensive influence efforts on truthful information and weigh the risks carefully.

Figure S.1
Defining the Terms and Scoping the Project

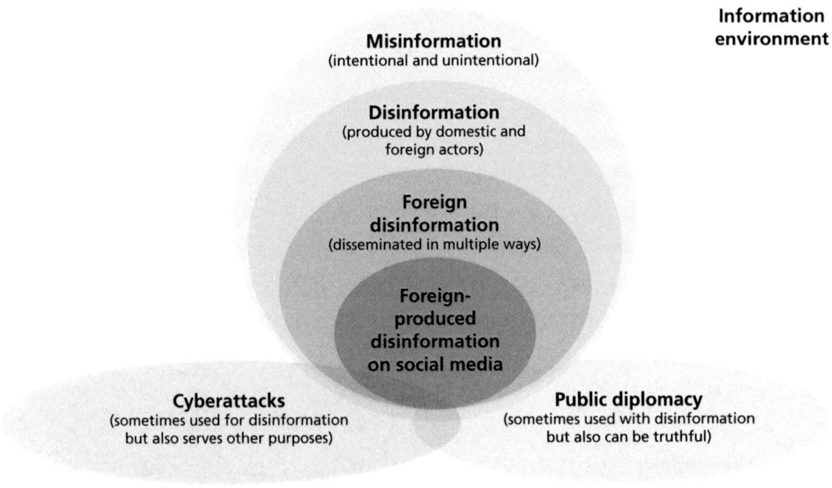

Acknowledgments

This report would not have been possible without the help of multiple people. First and foremost, we would like to thank Brig Gen Brenda Cartier, Maj Shaun Owenby, Capt Laura Jones and Air Force Special Operations Command for their support and direction throughout the course of the project. We would also like to thank Project AIR FORCE Strategy and Doctrine Program Director Paula Thornhill for her advice and guidance throughout the development of this work. We also owe a special debt of gratitude to the dozens of experts from across the public and private sectors whom we interviewed for the research for this report. While humans subjects protocols prevent us from acknowledging them by name, we would like to thank the following organizations: Atlantic Council, Bellingcat, Center for New American Security, Facebook, FireEye, Google, Graphika, Integrity Institute, JigSaw, Microsoft, National Endowment for Democracy, Johns Hopkins School of Advanced International Studies, Twitter, U.S. Department of Defense organizations: U.S. Air Forces–Europe, 67th Operational Support Squadron, Air Force OIE Tiger team, Air Staff Information Operations Branch, 151st Theater Information Operations Group (TIOG), Central Command Web Operations Division, Joint Staff J2, the Marine Corps Deputy Commandant for Information, II Marine Information Group, 1st Information Operations Command, Office of the Under Secretary of Defense, Policy, the U.S. Agency for Global Media, the Global Engagement Center, the Department of State's Global Public Affairs, the Department of State's Regional Bureaus, and the Congressional Research Service. Additionally, we would like to thank the dozens of academics, think tank analysts, military officers and government officials in Belarus, Belgium, Germany, Japan, Philippines, Singapore, South Korea, Taiwan, Ukraine, and the United Kingdom who were interviewed for this report. Although some of these organizations asked not to have their affiliations listed, we owe a tremendous debt of gratitude for their time and their insights.

Abbreviations

AFSOC	Air Force Special Operations Command
CENTCOM	Central Command
CYBERCOM	cyber command
DoD	U.S. Department of Defense
EU	European Union
FBI	Federal Bureau of Investigation
GEC	Global Engagement Center
GRU	Russian Military Intelligence
IRA	Internet Research Agency
IO	information operations
ION	Information Operations Network
MISO	Military Information Support Operations
NATO	North Atlantic Treaty Organization
NGO	nongovernmental organization
OIE	operations in the information environment
PLA	People's Liberation Army

PLAAF	PLA Air Force
PLASSF	PLA Strategic Support Force
RRN	Resident Registration Number
StratCom CoE	Strategic Communications Centre of Excellence
USAF	U.S. Air Force
USAGM	U.S. Agency for Global Media
USG	U.S. government
USIA	U.S. Information Agency
USMC	U.S. Marine Corps
VOA	Voice of America
VK	Vkontakte
WebOps	Web Operations Branch

CHAPTER ONE

Combating Foreign Influence Efforts on Social Media

Since the Russian interference in the 2016 American election, the question of how to best combat foreign disinformation has attracted significant public scrutiny and policy attention. Both the 2017 National Security Strategy and the 2018 National Defense Strategy identified "political and information subversion" and "influence campaigns [that] blend covert intelligence operations and false online personas with state-funded media, third-party intermediaries, and paid social media users or 'trolls'" as core threats to the U.S. homeland.[1] This report—part of a series sponsored by U.S. Air Force Special Operations Command (AFSOC)—examines these threats and the responses to date. It focuses on two sets of questions. First, how are state adversaries—particularly China and Russia—using disinformation on social media to advance their interests, and what does the joint force—and the U.S. Air Force (USAF) in particular[2]—need to be prepared to do in response? Second, what are the joint force, the U.S. government (USG), civil society, and private industry doing to combat these threats, how effective are these efforts, and what more needs to be done?

[1] White House, National Security Strategy of the United States of America, Washington, D.C., December 2017, p. 34; U.S. Department of Defense (DoD), Summary of the National Defense Strategy of the United States of America: Sharpening the American Military's Competitive Edge, Washington, D.C., 2018, p. 3.

[2] This report was completed before the creation of the U.S. Space Force and therefore uses the name "U.S. Air Force" to refer to both air and space capabilities.

This summary volume weaves the findings of the overall project together into a single holistic picture. Ultimately, this report presents mixed findings. Three years after the issue of foreign disinformation on social media became headline news, the response remains fractured, uncoordinated, and—by many actors' own admission—dubiously effective. At the same time, the threat of disinformation is far more nuanced than it is sometimes portrayed as being. Social media appears to be a far better medium for creating distrust and chaos than for building long-term influence. Russia and Iran have used this tool extensively; China and North Korea have used it in a more limited fashion. Even in the countries most affected by foreign disinformation campaigns, the effectiveness of these campaigns remain very much an open question.

Definitions and Methodology

What is disinformation? And what is social media? The answers are less straightforward than one might presume. There are multiple terms setting the stage for our discussion of disinformation and what DoD terms the *information environment*, or "the aggregate of individuals, organizations, and systems that collect, process, disseminate, or act on information."[3] In layman's terms, the information environment features a wide variety of actors—everyone from official media outlets to personal blog posts, from those who produce information (e.g., journalists) to those who read, watch, or listen to it. The vast majority of actors in the information environment are benign—at least from a national security perspective—and much of the information is truthful.

Within the subset of untruthful information, most of it can be characterized as *misinformation*—false, incomplete, or misleading information.[4] While misinformation is sometimes deliberately manu-

[3] DoD, *Department of Defense Dictionary of Military and Associated Terms*, Washington, D.C., June 2019.

[4] Merriam-Webster, "Misinformation," webpage, undated. Importantly, some definitions of misinformation deliberately exclude intentional efforts to mislead. For example, one handbook on the subject says, "Misinformation is generally used to refer to misleading information created or disseminated without manipulative or malicious intent." United Nations Educational,

factured, it does not need to be aimed at producing a specific effect. For any number of reasons, even well-intentioned, professional journalists sometimes get a story wrong.

By contrast, *disinformation* means "false, incomplete, or misleading information that is passed, fed, or confirmed to a target individual, group, or country."[5] Unlike misinformation, disinformation is designed to mislead a targeted population. Within a disinformation campaign, one or both of the following criteria apply:

- the information itself is false, incomplete or misleading
- the source of that information is false, incomplete, or misleading (e.g., a state actor passes itself off as private local individual).[6]

Multiple actors can produce disinformation. Political parties can produce disinformation to target their rivals; businesses might use disinformation to discredit their competitions; and foreign adversaries can use disinformation to undermine their adversaries.

Foreign disinformation can be spread through a variety of different media—television, radio, leaflets, and (most relevant to this study) social media. In their 2008 article, scholars Danah M. Boyd and Nicole B. Ellison defined *social media* as:

> Web-based services that allow individuals to (1) construct a public or semi-public profile within a bounded system, (2) articulate a list of other users with whom they share a connection, and

Scientific and Cultural Organization, *Journalism, 'Fake News' & Disinformation*, Paris, France, 2018, p. 7. Also see Catherine A. Theohary, *Information Warfare: Issues for Congress*, Washington, D.C.: Congressional Research Service, R45142, March 5, 2018, p. 5.

5 Richard H. Shultz and Roy Godson, *Dezinformatsiya: Active Measures in Soviet Strategy*, Washington, D.C.: Pergamon-Brassey, 1984, p. 41. For a similar definition, see United Nations Educational, Scientific and Cultural Organization, 2018, p. 7; Theohary, 2018, p. 5.

6 Of note, we also do not specifically look at the use of social media as an intelligence collection tool, although some adversaries have used social media in this capacity. For example, see Edward Wong, "How China Uses LinkedIn to Recruit Spies Abroad," *New York Times*, August 27, 2019.

(3) view and traverse their list of connections and those made by others within the system.[7]

In sum, our study's scope can be illustrated as a series of concentric rings with the study's primary focus—foreign disinformation on social media—depicted in dark orange at the center (Figure 1.1).

As depicted in Figure 1.1, two more categories are worth mentioning. First, cyberattacks can be conducted for any number of reasons—for example, to degrade an adversary and gather intelligence—including as part of disinformation campaigns, if hacked material is then manipulated and released to influence a targeted population. Second, all states conduct public diplomacy campaigns. The United States conducts *public diplomacy* with the expressed intent of "informing and influencing foreign publics and by expanding and strengthening the relationship between the people and Government of the United States and citizens

Figure 1.1
Defining the Terms

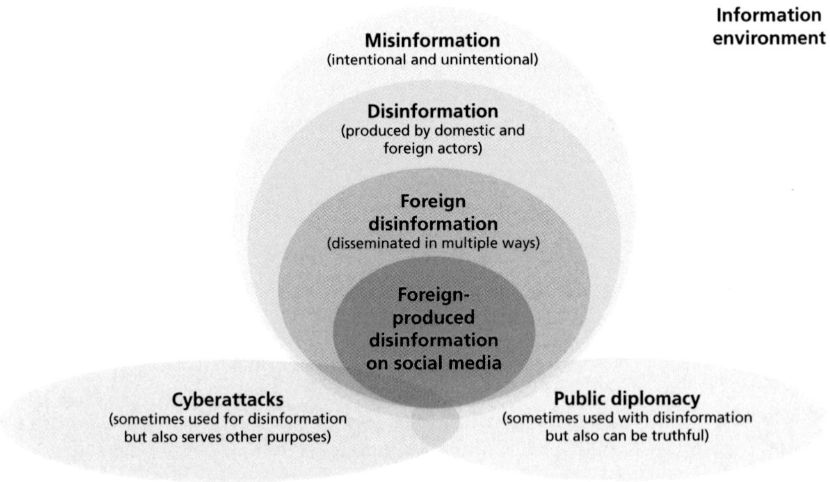

[7] Danah M. Boyd and Nicole B. Ellison, "Social Network Sites: Definition, History, and Scholarship," *Journal of Computer-Mediated Communication*, No. 13, 2008, p. 211.

of the rest of the world."[8] Public diplomacy is overt—clearly attributable back to the host government—and although the message does advance the government's point of view, it often claims to be truthful. Sometimes, however, these efforts are less than truthful, in which case this category can also overlap with disinformation efforts.

Admittedly, all these lines blur in practice. Foreign adversaries sometimes use domestic actors as part of their disinformation efforts. Similarly, in some parts of the world, large groups on direct messaging applications (such as WhatsApp) function in much the same way as more-traditional social media platforms (such as Facebook). Consequently, although we focused our research on the dark orange circle depicted in Figure 1.1, we admittedly stray from an overly doctrinaire application of this terminology.

Our research approach drew on a host of different primary and secondary sources, data sets, and (perhaps most importantly) more than 150 original interviews conducted around the world. Because multiple organizations—from the government to private sector—are involved in this issue, we attempted to interview to a representative cross section. To study the USAF response to disinformation on social media and how it fits with other public and private efforts, we interviewed a series of airmen from the tactical through the strategic levels. We also interviewed experts from sister services, the combatant commands, and joint staff to see how the USAF response integrates with that of DoD.

DoD, however, is only one of the government actors in this space. The Department of State also plays a role in combating disinformation abroad, so we interviewed individuals from the Public Affairs Bureau and the Global Engagement Center (GEC). We also talked to USG-affiliated organizations, such as Voice of America and National Endowment for Democracy, and government research institutions, such as the Congressional Research Service and National Defense University.[9]

---

8 U.S. Department of State, "Under Secretary for Public Diplomacy and Public Affairs: Our Mission," webpage, undated-b.

9 Of note, certain government agencies—such as the National Security Council Staff and parts of the intelligence community—declined to be interviewed for an unrestricted publication.

Unlike many more-traditional security threats, much of the first line of response to disinformation falls to the private sector. Therefore, the research team interviewed representatives from the major American social media companies, outside data analysis firms, and multiple universities and think tanks.

Finally, we also interviewed a mixture of foreign government officials, military officers, think tank analysts, and journalists about Russian, Chinese, and North Korean disinformation efforts. To understand Russian uses of disinformation—particularly in a tactical and operational context—we looked in depth at events in Ukraine and how Russia used disinformation in the context of the war in Donbass. Additionally, we interviewed several experts from across Europe—the United Kingdom, Belgium, Germany, and Belarus. For our investigation of Chinese disinformation, we also did a detailed analysis of Taiwan because, like Ukraine in the Russian context, Taiwan is arguably the test bed for Chinese disinformation tactics.[10] We also conducted interviews in Japan, Singapore, and the Philippines—all U.S. allies or partners, all home to U.S. military presences, and all plausible targets for Chinese disinformation, although interestingly, we uncovered less evidence of this activity than we anticipated. Lastly, we interviewed South Korean experts to gauge North Korean activity in this space.

This heavily interview-based research methodology allowed us to look broadly at how China and Russia are using this tool and how a variety of actors in the United States and around the world are responding to this threat. This approach did come with trade-offs, however. Most notably, we did not conduct original technical analyses to identify and expose disinformation, instead relying primarily on preexisting secondary sources.

The logic behind this trade-off was twofold. First, we aim to complement the more-technical analyses already published on these subjects. Second, an interview-heavy methodology seemed suited to

---

[10] China also allegedly used disinformation in response to the prodemocracy protests in Hong Kong, but this occurred after the bulk of the research for this report was complete.

the research questions focused more on understanding the threat eco-sphere, the policy response, and USAF's role in it.

There are limitations to this qualitative approach, especially when it comes to evaluating the effectiveness of disinformation campaigns. As will be discussed in Chapter Two, attributing results to any one disinformation campaign proves notoriously difficult, especially as we move from tactical uses of disinformation aimed at producing discrete effects to more-strategic disinformation campaigns aimed at shaping the attitudes of entire populations. "Proving" effectiveness, if it is possible, would likely require a more technical and more quantitative approach than presented here. Instead, we relied on the admittedly more-subjective qualitative expert evaluations combined with the findings of other, more-detailed analyses to evaluate effectiveness.

Organization and the Argument

In total, this work consists of three other volumes. Two volumes in the series—*Chinese Disinformation Efforts on Social Media* and *Russian Disinformation Efforts on Social Media*—look at how China and Russia have used disinformation on social media in the past, how they will likely do so in the future, and what the United States, the joint force, and USAF in particular must consider in response.[11] The other volume, not available to the general public, examines how the USG, DoD, and USAF combat disinformation on social media and train for its use in conflict.[12] Moreover, because social media platforms are run by private companies, much of the response to disinformation falls

---

[11] Scott W. Harold, Nathan Beauchamp-Mustafaga, and Jeffrey W. Hornung, *Chinese Disinformation Efforts on Social Media*, Santa Monica, Calif.: RAND Corporation, RR-4373/3-AF, 2021; Elina Treyger, Joe Cheravitch, and Raphael S. Cohen, *Russian Disinformation Efforts on Social Media*, Santa Monica, Calif.: RAND Corporation, RR-4373/2-AF, forthcoming.

[12] Raphael S. Cohen, Alyssa Demus, Michael Schwille, and Nathan Vest, *U.S. Efforts to Combat Foreign Disinformation Efforts on Social Media*, Santa Monica, Calif.: RAND Corporation, 2021, Not available to the general public.

on the private sector, so this volume also looks at how industry has attempted to police these platforms.

The remainder of this report draws on the other volumes' findings to provide an overview of the threat of disinformation, the response, and the recommendations for how best to combat it in the future—particularly in terms of USAF's equities. In Chapter Two, we draw on the analyses of Russian, Chinese, North Korean, and (to a degree) Iranian disinformation efforts on social media to describe the threat environment. In Chapter Three, we provide an overview of how industry, the USG, DoD, and some partners and allies are responding to the threat. Finally, we conclude with recommendations for both USAF and the USG.

Ultimately, this report presents mixed findings. For all the hyperbole surrounding disinformation efforts on social media, the actual threat is more nuanced than it is sometimes portrayed. Russia has been quite active in this space; China and North Korea less so internationally (although they are active in the disinformation space directed toward their own populations). Russia has racked up clear operational successes—sowing chaos and exacerbating preexisting social cleavages—but it remains unclear whether any of these campaigns on social media have had a long-term, strategic impact on any of their targets.

Three years after the 2016 elections, the USG, the joint force, civil society, tech companies, and U.S. allies and partners around the world are still grappling with how to confront the threat of disinformation on social media. The only point of consensus among more than 150 experts interviewed for this study is that the response to date remains fractured and ad hoc, and no solution has been found.

CHAPTER TWO

Understanding the Threat

Disinformation campaigns on social media are not a monolithic threat. Rather, disinformation is a long-standing tool of statecraft now applied to the fairly new realm of social media to create a new variant of information warfare that different states have employed in different ways—not unlike when the machine gun was added to aircraft to create a new realm of aerial warfare a century ago. Consequently, despite all the attention given to the subject, there is still much we do not know about social media, including how effective it is in producing strategic outcomes. What we do know is that China, Russia, North Korea, and Iran have all chosen to wield this tool in this domain very differently in the service of a variety of objective ends. Consequently, the threats that the United States and its allies face from adversaries' activities look very different across the globe.

A Tool of the Weak?

In July 2018, Homeland Security Secretary Kirstjen Nielsen warned of "persistent Russian efforts using social media, sympathetic spokespeople and other fronts to sow discord and divisiveness amongst the American people, though not necessarily focused on specific politicians or political campaigns."[1] Analysts aptly describe Russia engaging in a "firehose of falsehoods" because of its high-volume multi-

---

1 Geoff Mulvihill, "US Official: Russia Using Social Media to Divide Americans," Associated Press, July 15, 2018.

channel approach to its propaganda.[2] The United States might be a high-priority target, but Russia certainly has conducted disinformation efforts elsewhere. Across Europe, there are reports of Russian disinformation efforts to interfere in the social fabric and the political inner workings of states.[3] Extrapolating from the Russian example, one can easily be led to believe the world is awash in foreign disinformation efforts, but this would be incorrect: There is plenty of domestically produced disinformation, but there is less clear evidence of states mounting coordinated disinformation campaigns against rival actors on the scale and scope of Russia's actions.

Iran—though not an explicit focus of this study—perhaps comes closest. In 2018, researchers uncovered a large Iranian disinformation campaign sprawling across the United States, Europe, Latin America, and the Middle East that tried to promote anti-Saudi, anti-Israeli, pro-Palestinian, and pro-Iranian narratives.[4] A Reuters-led study found another far-flung Iranian disinformation attempt stretching from Indonesia to Sudan to the United Kingdom and the United States.[5] In May 2019, Twitter took down 2,800 Iranian-linked fake accounts, including some that falsely claimed to represent political candidates, as part of a larger crackdown on Iranian influence campaigns.[6]

---

[2] Christopher Paul and Miriam Matthews, *The Russian "Firehose of Falsehood" Propaganda Model: Why It Might Work and Options to Counter It*, Santa Monica, Calif.: RAND Corporation, PE-198-OSD, 2016.

[3] Raphael S. Cohen and Andrew Radin, *Russia's Hostile Measures in Europe: Understanding the Threat*, Santa Monica, Calif.: RAND Corporation, RR-1793-A, 2019; Todd C. Helmus, Elizabeth Bodine-Baron, Andrew Radin, Madeline Magnuson, Joshua Mendelsohn, William Marcellino, Andriy Bega, and Zev Winkelman, *Russian Social Media Influence: Understanding Russian Propaganda in Eastern Europe*, Santa Monica, Calif.: RAND Corporation, RR-2237-OSD, 2018.

[4] FireEye Intelligence, "Suspected Iranian Influence Operation Leverages Network of Inauthentic News Sites & Social Media Targeting Audiences in U.S., UK, Latin America, Middle East," blog post, August 21, 2018.

[5] Jack Stubbs and Christopher Bing, "Special Report: How Iran Spreads Disinformation Around the World," Reuters, November 30, 2018.

[6] Tony Romm, "Facebook and Twitter Disable New Disinformation Campaign with Ties to Iran," *Washington Post*, May 29, 2019.

Still, according to publicly available information, North Korea seems to have used social media more as an auxiliary tool for cyber-espionage than for disinformation. In 2018, the U.S. Department of Justice indicted Park Jin Hyok, a programmer believed to have participated in the 2014 cyberattack on Sony Pictures Entertainment.[7] According to the affidavit, Korean hackers made extensive use of fake Facebook, Twitter, and LinkedIn personas—such as "Andoson David," "Watson Henny," and "John Mogabe"—to conduct initial reconnaissance of Sony Pictures Entertainment affiliates and to spread malware by posting comments containing malicious links on various actors' Facebook posts.[8]

To an even greater extent, North Korea has also used social media to target South Korean audiences. According to a 2016 National Assembly briefing by South Korea's National Intelligence Service, North Korea created a fake Facebook account with a profile picture of a beautiful young woman and sent friend requests to dozens of South Korean government officials asking for sensitive information.[9] An official at Korea Internet & Security Agency confirmed later in the same year that North Korea has used such tactics.[10]

The real enigma here, perhaps, is China. On the surface, China should be a key player in foreign influence campaigns on social media. China has a long-standing interest in political warfare. The People's Liberation Army (PLA)'s well-known "three warfares" strategy combines psychological, public opinion, and legal warfare.[11] China's national military guideline (the equivalent of its military doctrine)

---

7 *United States of America v. Park Jin Hyok*, U.S. District Court for the Central District of California, June 8, 2018.

8 *United States of America v. Park Jin Hyok*, 2018.

9 Kang Tae-Hwa [강태화], "North Korea 'Facebook Honey Trap' . . . Befriended Officials to Ask for Documents [강태화 <북한 '페북 미인계>…공직자와 친구 맺어 자료 요구]," *JoongAng Ilbo* [중앙일보], 2016.

10 Lee Hyo-Suk [이효석], "Beautiful Facebook Friend May Be A Spy . . . North Korea's Cyber Terrorism Diversifies [미모의 페친, 알고보니 간첩일수도…북 사이버테러 다양화]," Yonhap News Agency [연합뉴스], 2016.

11 For an overview of the three warfares, see Peter Mattis, "China's 'Three Warfares' in Perspective," War on the Rocks, January 30, 2018.

and a 2015 defense white paper both focus on the need to prepare to fight "informationized local wars" (信息化局部┌┌).[12] Moreover, China reportedly spends perhaps $10 billion a year on propaganda; even a small portion of that going toward social media would have a large impact.[13] A Taiwan government official said that the PLA Strategic Support Force (PLASSF)—the arm of the PLA tasked with information warfare, among other functions—had at least 175,000 troops.[14]

According to open source reporting, China has not been nearly as active in terms of mounting disinformation campaigns abroad. Much of the Chinese government's effort on propaganda, including social media, is focused domestically. Studies suggest that somewhere between 0.6 percent and 16.7 percent of all domestic posts are from accounts affiliated in some way with the Chinese Communist Party.[15] Looking abroad, one study found that "the selected public accounts run by *Xinhua News*, *People's Daily*, and CCTV News/CGTN have established a significant presence in the Twittersphere in the six-and-a-half years or so since they started their accounts," though they still trailed Russia's *RT*.[16] Another study examined *People's Daily* and *Xinhua* on Instagram and found, "These two Chinese influence profiles reached a level of audience engagement roughly one-sixth as large as the entire

---

[12] Chinese State Council Information Office, China's Military Strategy, via *Xinhua*, May 2015.

[13] David Shambaugh, "China's Soft-Power Push," *Foreign Affairs*, July 2015.

[14] International Institute for Strategic Studies, *Military Balance 2019*, London, February 2019, p. 262. Of note, some place the strength of PLASSF at considerably larger figures. See Jason Pan, "China Subverting Elections: Premier," *Taipei Times*, November 2, 2018.

[15] Gary King, Jennifer Pan, and Margaret E. Roberts, "How the Chinese Government Fabricates Social Media Posts for Strategic Distraction, Not Engaged Argument," *American Political Science Review*, Vol. 111, No. 3, 2017; Mary Gallagher and Blake Miller, "Can the Chinese Government Really Control the Internet? We Found Cracks in the Great Firewall," *Washington Post*, February 21, 2017.

[16] Joyce Y. M. Nip and Chao Sun, "China's News Media Tweeting, Competing With US Sources," *Westminster Papers in Communication and Culture*, Vol. 13, No. 1, 2018.

Russian Internet Research Agency (IRA)–associated campaign targeting the United States on Instagram."[17]

Our study found relatively few Chinese disinformation campaigns along the lines of the Russian model, an exception being China's information efforts in Taiwan, where China is quite active. One interviewee from Taiwan claimed that China is attacking Taiwan with as many as 2,400 separate pieces of disinformation every day.[18] As is detailed in our report on China in this series, Taiwan officials can give dozens of examples of what they claim are Chinese-linked disinformation narratives.[19]

More recently, Twitter, Facebook, and YouTube announced the discovery of an extensive disinformation effort tied to creating division within and discrediting the pro-democracy Hong Kong protests. As of August 22, 2019, YouTube had suspended at least 210 accounts; Twitter had inactivated some 936 accounts with 200,000 more under suspicion; and Facebook had disabled five accounts, seven pages and three groups—all believed to be tied to Chinese disinformation campaigns.[20] Some of these accounts featured content in Chinese; others were in English; and at least one of the Twitter accounts (@LibertyLionNews) seemed to target a U.S. audience.[21]

To date, however, Taiwan and Hong Kong seem to be the exception among China's disinformation efforts. Field research in Japan, Singapore, and the Philippines unearthed relatively few clear-cut cases of Chinese disinformation. There is plenty of evidence of other forms of Chinese influence—public diplomacy efforts, investments in key

---

17 Insikt Group, *Beyond Hybrid War: How China Exploits Social Media to Sway American Opinion*. Boston, Mass.: Recorded Future, 2019.

18 Interview with Chinese disinformation researcher, Taipei, Taiwan, January 2019.

19 Harold, Beauchamp-Mustafaga, and Hornung, 2021.

20 Ursula Perano, "YouTube Disables 210 Channels Linked to Hong Kong Influence Campaign," *Axios*, August 22, 2019; Twitter Safety, "Information Operations Directed at Hong Kong," Twitter Blog, August 19, 2019; Nathaniel Gleicher, "Removing Coordinated Inauthentic Behavior From China," Facebook Newsroom, August 19, 2019; Donie O'Sullivan, "How a Hacked American Nightclub Twitter Account Was Implicated in China's Information War," CNN, August 21, 2019.

21 Gleicher, 2019; Perano, 2019; Twitter Safety, 2019.

industries and locations, and cultural outreach—but not of disinformation efforts on social media per se. This is not for lack of motivation or opportunity: Japan and the Philippines are U.S. allies, and Singapore is a key staging point for the U.S. military. This makes them attractive targets for a rising China looking to expand and cement its reach over the Indo-Pacific, and all three states have cleavages that China theoretically could exploit through disinformation on social media if it chose to do so. Singapore is a multiethnic society; the Philippines has economic, religious and ethnic divides and, recently, a precarious relationship with the United States; and Japan already has naturally occurring resentment toward U.S. basing in Okinawa. Ultimately, Chinese disinformation is almost more striking for its absence than its presence.

There are many possible reasons why the Chinese have not invested in disinformation campaigns on social media to the same degree that the Russians have. First, it is possible that China is actively conducting foreign disinformation campaigns but is better at tradecraft, so these campaigns have yet to be detected. This hypothesis is not entirely compelling, however. Taiwan has clearly been exposed repeatedly to Chinese disinformation efforts, and the social media giants, such as Facebook and Twitter, detected Chinese disinformation efforts surrounding the Hong Kong protests, so it appears that Chinese disinformation efforts are not entirely clandestine.

Another reason might be that China is focused on shielding itself from foreign disinformation campaigns before developing the capability to conduct these efforts abroad. The PLASSF was established in 2015 and is "accelerat[ing] China's use of disinformation significantly."[22] If this is the case, then China could become more active in this arena in the future. However, this explanation is not particularly compelling, either. A country as large and as wealthy as China should have the capacity to build whatever protections it feels it needs to defend against disinformation while also developing an offensive capability—if it chose to do so.

Yet another possibility is that China—a wealthy, aspiring hegemonic power with multiple tools of influence at its disposal—does not

---

[22] Interview with Chinese disinformation researcher, Taipei, Taiwan, January 2019.

view stirring up social discord through social media or other means to be in its interest. China has certainly developed the capability to do so (as it has demonstrated in Taiwan) and will employ it elsewhere under specific circumstances, particularly to discredit and divide opposition to the regime (as it demonstrated in the Hong Kong protests). That said, China might view disinformation as a path to sustainable long-term influence. Arguably, if China is interested in building a more China-friendly order in Asia, then expanding and cementing Chinese influence through political, financial, and cultural ties might be a more prudent policy option than spreading disinformation on social media and simply creating chaos.

In other words, there is a plausible case that disinformation campaigns on social media might be a weapon of the weak. As we shall show in the next section, it is a means far better suited for sowing bedlam rather than building long-term influence. It might be mostly employed rogue states, not peer competitors, that can reasonably aspire to build a lasting order.[23]

A New Frontier Still In Its Infancy

One of the challenges of understanding why China is not as active in disinformation campaigns on social media is because the tool itself is still in its infancy.[24] Multiple powers—including China, Russia, and Iran—have long histories of employing disinformation as part of a broader array of measures short of war—or as famed Cold War diplomat and historian George Kennan referred to it—*political warfare*.[25]

---

23 For a similar argument, see James Dobbins, Howard J. Shatz, and Ali Wyne, *Russia Is a Rogue, Not a Peer; China Is a Peer, Not a Rogue: Different Challenges, Different Responses*, Santa Monica, Calif.: RAND Corporation, PE-310-A, 2019.

24 For a similar conclusion, see Michael J. Mazarr, Abigail Casey, Alyssa Demus, Scott W. Harold, Luke J. Matthews, Nathan Beauchamp-Mustafaga, and James Sladden, *Hostile Social Manipulation: Present Realities and Emerging Trends*, Santa Monica, Calif.: RAND Corporation, RR-2713-OSD, 2019.

25 For a brief history and overview of these countries' efforts, see Linda Robinson, Todd C. Helmus, Raphael S. Cohen, Alireza Nader, Andrew Radin, Madeline Magnuson, and Katya

That said, disinformation on social media is a rather recent phenomena, partly because social media itself is fairly new.

Social media's precursors date back to the Bulletin Board Systems of the 1980s, but social media in its modern incarnation largely began with Friendster and Myspace, launched in 2002 and 2003, respectively.[26] Since then, social media has grown at a meteoric rate—faster than population growth or global internet penetration—and by 2018, more than 2.6 billion people globally used social media (see Figure 2.1).

The relatively recent explosion of social media has three key implications for foreign disinformation campaigns. First, despite the fact that

Figure 2.1
Year-over-Year Growth in Internet and Social Media Users

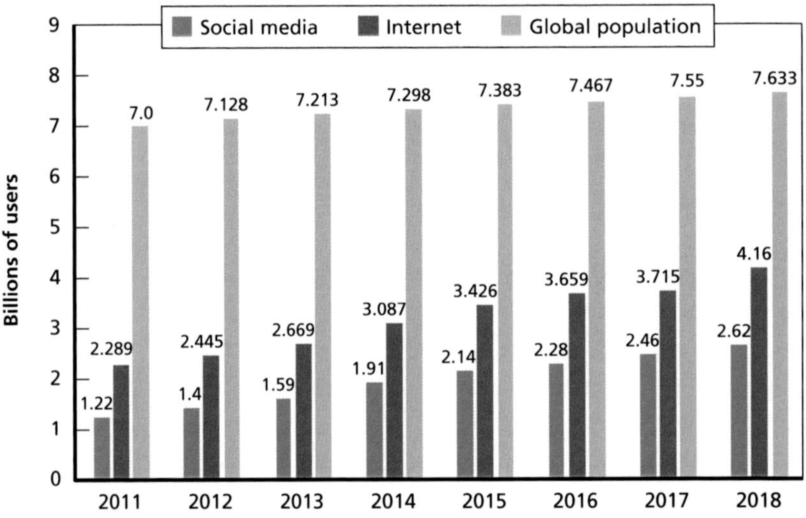

SOURCES: Statista, "Internet Penetration Rate in the Middle East Compared to the Global Internet Penetration Rate from 2009 to 2018," webpage, March 2018b; Statista, "Number of Social Media Users Worldwide from 2010 to 2021 (in billions)," webpage, May 2018d; Worldometer, "Current World Population," webpage, undated-a.
NOTE: The world population was 7,691,587,976, as of March 20, 2019.

Migacheva, *Modern Political Warfare: Current Practices and Possible Responses*, Santa Monica, Calif.: RAND Corporation, RR-1772-A, 2018.

[26] Beata Biały, "Social Media—From Social Exchange to Battlefield," *Cyber Defense Review*, Vol. 2, No. 2, Summer 2017, p. 69.

all of the U.S.-named adversaries have deep roots in using disinformation, not all of them embraced disinformation on social media as a tool at the same rate. Russia seemingly was one of the first to realize the power of disinformation on social media as an offensive weapon. Russian military writings suggest that Moscow first perceived the offensive implications of emerging communications technology in the 1990s and more fully embraced the power of social media in the early 2000s in response to what it believed was the United States' growing dominance in this domain. Russian military experts paid close attention to all technologies used for psychological operations: For example, a Russian military psychological operations officer was impressed with the North Atlantic Treath Organization (NATO)'s use of the Commando-Solo broadcasting platform and with the 193rd Air Wing in Yugoslavia, which supplemented an online NATO propaganda effort allegedly involving "more than 300,000 websites."[27]

Russia's online disinformation tactics started domestically. During the Chechen conflict in the late 1990s, both Russian state and pro-Russian nonstate actors attacked Chechen online media and other websites—which, although best described as hacking, appear to have aimed at "informational-psychological" effects.[28] Later, Russia began to apply these lessons to foreign campaigns. Drawing lessons from the 2008 war with Georgia, a Russian colonel noted South Ossetia's success in organizing mass influence efforts through the internet, as it turned to blogs and social media to counter Georgian messaging. [29] Such "mass information armies," the author argued, were more effec-

---

[27] The EC-130 Commando-Solo is a modified transport aircraft that can broadcast messages on radio and television. As such, it is a key delivery platform for military information efforts. Vladimir Akhmadullin, "The Word, Equal to the Bomb [Слово, приравненное к бомбе]," *Independent Military Review [Независимое военное обозрение]*, No. 25, July 2, 1999.

[28] For example, see Daniil Turovsky, "'It's Our Time to Serve the Motherland': How Russia's War in Georgia Sparked Moscow's Modern-Day Recruitment of Criminal Hackers," *Meduza*, August 7, 2018.

[29] P. Kolesov, "Georgia's Information War Against South Ossetia and Abkhazia [Информационная война Грузии против Южной Осетии и Абхазии]," *Foreign Military Review [Зарубежное военное обозрение]*, No. 10, October 2008.

tive than the "mediated" dialogue of state leaders with the peoples of the world.[30] By 2014, as the Maidan uprising deposed pro-Russian Ukrainian president Viktor Yanukovich, Russia was ready to deploy this new technology alongside kinetic operations.

China, by contrast, appears to have moved considerably more slowly. The Chinese government's first foreign social media account was opened on Twitter in 2009 by China Radio International (now China Plus News); by 2012, most of China's main state-run media had accounts, including *China Daily*, *Xinhua*, *People's Daily* and *Global Times*.[31] This relatively slow adoption of the foreign communication technologies fits a broader pattern of Chinese behavior: China launched its first foreign TV channel in 1992 and its first foreign website in 1997.[32]

As for the military, there are articles in PLA propaganda journals dating as far back as 2014 arguing for establishing a presence on Facebook, Twitter, and other platforms. But the PLA, at least so far, eschews foreign social media. [33] The PLA Air Force (PLAAF) opened the first PLA account on Weibo and WeChat in October 2015, and now has almost 2.5 million followers on Weibo, more than any other service.[34] It does not have a presence on foreign social media platforms, however, and much of the PLAAF's content on Weibo is clearly intended to garner support for the PLAAF and might be designed, at least in part, for recruitment purposes.[35]

---

[30] Kolesov, 2008.

[31] For a review of China's international social media engagement, see Nip and Sun, 2018.

[32] Nip and Sun, 2018.

[33] Nathan Beauchamp-Mustafaga and Michael Chase, *Borrowing a Boat Out to Sea: The Chinese Military's Use of Social Media for Influence Operations*, Washington, D.C.: John Hopkins School of Advanced International Studies, 2019.

[34] Follower count as of June 15, 2019. See PLAAF, "Air Force Release [空┌┌布]," Weibo, undated; Ministry of National Defense of the People's Republic of China, "Chinese Air Force Official: Weibo, WeChat Public Account Open [中┌空┌官方微博、微信公┌┌┌通┌行]," press release, November 10, 2015.

[35] Derek Grossman, Nathan Beauchamp-Mustafaga, Logan Ma, and Michael Chase, *China's Long-Range Bomber Flights: Drivers and Implications*, Santa Monica, Calif.: RAND Corporation, RR-2567-AF, 2018.

Accusations of Chinese disinformation on foreign social media first began in Taiwan following President Tsai Ing-Wen's election in 2016. Many of the examples from mid-2016 onward focused on undermining support for Tsai by claiming she was mismanaging the military or damaging Taiwan's traditional culture.[36] One report in 2014 identified fake Twitter accounts (bots) that were broadcasting positive messages about Tibet, suggesting at least some parts of the Chinese government had a covert presence on foreign social media before 2016.[37] Moreover, reports emerged in 2015 and again in 2016 that state-run media accounts on Twitter were buying followers as a way to artificially increase their influence, suggesting interest in covert tactics before 2016.[38] Chinese intelligence also has been accused of using social media for recruitment several times since 2017, an indication that the Chinese government sees value in social media beyond simple propaganda.[39] If all these reports are true, it suggests that China waded into the social media space only over the past five years or so.

Iran and North Korea are also both relatively new to the business of disinformation on social media. Iran became interested in disinformation campaigns on social media at least as early as Russia. The cybersecurity firm FireEye identified fake Twitter accounts dating to 2011 that were designed to promote pro-Iranian policies.[40] North

---

36 J. Michael Cole, "Fake News at Work: President Tsai 'Persecutes Religion in Taiwan,'" Taiwan Sentinel, July 20, 2017; Lu Hsin-hui, Hsieh Chia-chen, Yeh Tzu-kang, and Elizabeth Hsu, "Authorities Deny Rumor of Ban on Incense, Ghost Money Burning," *FocusTaiwan*, July 21, 2017.

37 Jonathan Kaiman, "Free Tibet Exposes Fake Twitter Accounts by China Propagandists," *The Guardian*, July 22, 2014.

38 Tom Grundy, "Did China's State-Run News Agency Purchase Twitter Followers?" *Hong Kong Free Press*, April 14, 2015; Alexa Olesen, "Where Did Chinese State Media Get All Those Facebook Followers?" *Foreign Policy*, July 7, 2015; Nicholas Confessore, Gabriel J. X. Dance, Richard Harris, and Mark Hansen, "The Follower Factory," *New York Times*, January 27, 2018.

39 "German Spy Agency Warns of Chinese LinkedIn Espionage," BBC News, December 10, 2017.

40 Alice Revelli and Lee Foster, "Network of Social Media Accounts Impersonates U.S. Political Candidates, Leverages U.S. and Israeli Media in Support of Iranian Interests," FireEye Threat Research, blog post, May 28, 2019.

Korea's efforts are hard to date, but the February 26, 2014, edition of *Rodong Sinmun* reported that Kim Jong-Un stated at a large-scale Workers Party of Korea event:

> [We] must establish critical measures to make the Internet a place for propaganda for our thoughts and culture, in response to the imperialists' extensive dissemination of reactionary material using appropriated latest technology developed by humanity. . . . [We] must establish a plan to modernize and informatize means of domestic and foreign propaganda in areas of thought operations and related units. [41]

Kim's exact intention behind this statement is unknown, but it is probably safe to assume that North Korea has been interested in conducting disinformation campaigns on social media at least since the time that this speech was given—and possibly before.

The second major implication of the relative newness of social media and foreign disinformation campaigns on social media is that the effort to combat these campaigns is even more nascent. As we will detail in the next chapter, the United States has struggled at the interagency level and within the military over how best to respond to foreign disinformation on social media during peacetime and how to prepare for its potential use in a more overt military conflict. The United States is not alone in this respect. Around the world, governments are experimenting with new approaches—from new laws to media education programs—all of which are still in their infancy.

The advent of foreign disinformation on social media has also thrust private industry to the forefront of geopolitics and information warfare in ways never seen before. In previous generations, the state at least regulated and sometimes directly owned large parts of print, radio, or even television outlets and thus could exert control over the information space. By contrast, social media companies are private

---

41 "Advancing the Final Victory with a Revolutionary Offensive Offensive: Speech by Dear Kim Jong-Un at the 8th Annual Military Conference of the Workers' Party of Korea [혁명적인 사상공세로 최후승리를 앞당겨나가자: 경애하는 김정은 동지께서 조선로동당 제8차 사상일군대회에서 하신 연설]," *Labor News* [*Rodong Sinmun,* 노동신문], 2014.

entities, often operating across state boundaries. To further complicate matters, unlike other traditional media outlets, social media platforms' content is generated by users rather than by the companies. As a result, even if these companies want to prevent disinformation, they must first identify it—a nontrivial challenge considering that Facebook alone had some 2.32 billion users in December 2018.[42]

Finally, the third major implication of the newness of disinformation campaigns on social media is that this still very much an evolving threat. Unlike disinformation transmitted via other media—print, radio, or television—disinformation on social media opens new frontiers for personalization and microtargeting. By its nature, social media is tailored to individuals based on their preferences, which means that disinformation can also be tailored to the individual. As the ability to gather and use personal information becomes increasingly sophisticated, the scope of what is possible in disinformation campaigns will also expand. Any study that assess at the relative balance of offensive versus defense balance in online disinformation, including this one, must be viewed as a snapshot in time rather than as an everlasting truth.

Intimidate, Discredit, Divide

Obama administration Deputy National Security Advisor and Deputy Secretary of State Anthony Blinken was often fond of saying that "superpowers don't bluff."[43] The statement as description of historical fact is, of course, false: Great powers—including the United States—make false or misleading statements for a variety of reasons. As policy prescription, however, the logic supporting Blinken's statement is arguably sound. A kernel of truth lies at the core of many of the most-successful disinformation campaigns; disinformation alone can rarely persuade individuals to adopt new ideas wholesale, so these campaigns

[42] Facebook Newsroom, "Company Info," webpage, undated.

[43] Jonathan Allen, "Tony Blinken's Star Turn," *Politico*, September 16, 2019.

usually need some factual foundation. From that foundation, disinformation can be crafted to achieve various goals.

First, it can be used as a tool for intimidation. China, for example, propagated an image of a Chinese H6K bomber flying near Taiwan's iconic Jade Mountain, presumably with the intent of reinforcing the ideas of China's military supremacy, Taiwan's defense inadequacies, and the futility of resistance.[44] After the photos were released, Taiwan's Ministry of Defense denied the planes flew close enough to take such a photo, strongly suggesting it was disinformation.[45] The Taiwan spokesperson said the release

> is a typical act of propaganda [employed by China], and the [Taiwan] media are helping China in its 'advertising campaign' . . . . The goal [of the photographs' release] is to affect Taiwanese psychologically. There will probably be another picture released tomorrow, as China is thrilled with the reaction of the Taiwanese media.[46]

Sure enough, on February 3, 2019, the PLAAF released a video titled "Our Fighting Eagles Fly Circles around Taiwan."[47]

Russia, similarly, used disinformation sent via SMS (short message service) text messages to intimidate Ukrainian soldiers during the height of the war in Eastern Ukraine.[48] Using personal data that were possibly harvested from the soldiers' social media accounts, Russia

---

[44] "PLA Air Force Releases Apparent H-6K Photographed with Taiwan's Jade Mountain [解放┌空┌┌布疑似┌−6K┌台┌玉山合影]," *Observer* [┌察者], December 17, 2016.

[45] Matthew Strong, "Military Denies Yushan in China Bomber Picture: Peak Likely to Be Mount Beidawu in Southern Taiwan: Experts," *Taiwan News*, December 17, 2016.

[46] Chen Wei-han, "MND Plays Down China Aircraft Threat," *Liberty Times*, December 19, 2016.

[47] Aaron Tu and William Hetherington, "Defense Bureau to Tackle Propaganda from China," *Taipei Times*, March 4, 2019.

[48] Helmus et al., 2018, p. 16; Emilio J. Iasiello, "Russia's Improved Information Operations: From Georgia to Crimea," *Parameters*, Vol. 47, No. 2, 2017, p. 55; Aaron Brantly and Liam Collins, "A Bear of a Problem: Russian Special Forces Perfecting Their Cyber Capabilities," Association of the United States Army, November 28, 2018.

managed to personalize its disinformation effort via electronic targeting.[49] Soldiers reported receiving a series of text messages saying such things as, "[Y]ou are about to die. Go home," or "[T]his is not your war, this is the oligarchs' war, your family is waiting for you."[50] In some cases, the messages were made to look like they came from a soldier's relative.[51] Family members of Ukrainian soldiers also reported receiving personalized messages, presumably in a similar effort to intimidate them—and, by extension, their loved ones serving at the front.[52]

Disinformation also can be used to discredit an adversary. For example, when Typhoon Jebi hit Osaka, Japan, the Chinese allegedly planted a story on the private Taiwan-based social media platform PTT saying that the director of the Taipei Economic and Cultural Representative Office (Taiwan's unofficial vehicle for managing bilateral issues) did nothing to help stranded Taiwan citizens while the Chinese consulate in Osaka had dispatched buses to rescue trapped citizens.[53] This fake news story, part of a broader effort to discredit Taiwan's government, had a real-world impact: The accused government official committed suicide after coming under intense criticism online.[54]

Russia uses disinformation as a key tool for discrediting evidence of its own misdeeds. According to Russia expert Mark Galeotti, "[t]he next best thing to being able to convince people of your argument, after all, is to make them disbelieve all arguments."[55] For exam-

---

[49] Interview with journalist, Kyiv, Ukraine, March 5, 2019.

[50] Interview with journalist, Kyiv, Ukraine, March 5, 2019; interview with security officials, Kyiv, Ukraine, March 6, 2019; Brantly and Collins, 2018; Nataliia Popovych and Oleksiy Makhuhin "Countering Disinformation: Some Lessons Learnt by Ukraine Crisis Media Center," Ukraine Crisis Media Center, April 20, 2018.

[51] Interview with security officials, Kyiv, Ukraine, March 6, 2019.

[52] Interview with Ukraine media expert, Kyiv, Ukraine, March 8, 2019.

[53] Ko Tin-yau, "How Fake News Led to Suicide of Taiwan Representative in Osaka," *EJInsight*, September 19, 2018.

[54] Ko Tin-yau, 2018. The attribution back to China is uncertain. Other interviews in Taipei suggested that Taiwan students were the source of this allegation—not China.

[55] Mark Galeotti, *Controlling Chaos: How Russia Manages Its Political War in Europe*, London: European Council of Foreign Relations, 2017, p. 6.

ple, when Russia was caught trying to assassinate Sergei Skripal, a former Russian intelligence agent living in the United Kingdom, it tried to discredit the news through disinformation. As of early 2019, the European Union (EU)'s East StratCom Task Force counted more than 40 different accounts for the Skripal poisoning.[56] Russian actors opted for quantity over quality, apparently aiming to dominate social media conversations pertaining to these actions. The Atlantic Council's Digital Forensics Lab analysis showed that over the course of a week in 2018, two out of three articles on the poisoning shared via four key social media platforms—Facebook, Twitter, LinkedIn, and Pinterest—"came from Kremlin-funded media outlets."[57] These operations likely aimed to create an impression that truth cannot be ascertained, and only various versions of events exist.[58]

Like Russia, North Korea has used disinformation to deny its own wrongdoings to maintain plausible deniability and undermine justification for retaliatory measures. After sinking the South Korean corvette *Cheonan* in 2010, North Korea launched a sophisticated disinformation campaign that sought to deny its involvement in the attack, undermine South Korea's retaliatory sanctions, and generate sympathetic public opinion within South Korea that North Korea was being scapegoated. North Korean operatives used South Korean citizens' Resident Registration Numbers (RRNs) to create fake accounts with online forums that then posted messages similar to those posted on overt channels.[59] The appropriated RRNs were later found to belong to

[56] EU vs. Disinfo, "Year in Review: 1001 Messages of Pro-Kremlin Disinformation," webpage, January 3, 2019.

[57] *Donara Barojan*, "#PutinAtWar: Social Media Surge on Skripal," *Medium*, April 5, 2018a.

[58] For example, see Robinson et al., 2018, p. 65; Jean-Baptiste Jeangène Vilmer, Alexandre Escorcia, Marine Guillaume, and Janaina Herrera, *Information Manipulation: A Challenge for Our Democracies*, Paris: Policy Planning Staff (CAPS) of the Ministry for Europe and Foreign Affairs and the Institute for Strategic Research (IRSEM) of the Ministry for the Armed Forces, August 2018, p. 75.

[59] Lee Kui-Won [이귀원], "North Korea Appropriated South Korean Resident Registration Number (RRN) to Spread Rumors About Sinking of the Cheonan [북, 주민번호 도용 '천안함 날조' 유포]," Yonhap News Agency [연합뉴스], 2010.

regular South Korean citizens, including children in elementary school and housewives.[60]

Finally, disinformation can be used to divide and generally cause chaos. Interviewees from Taiwan noted that China seems particularly interested in exacerbating social divisions by targeting niche groups inside Taiwan's society—young people, retired Republic of China military officers, pensioners, religious groups within Taiwan society, farmers and fishermen, and those deeply attached to one political party or the other.[61] In the summer of 2017, a rumor ultimately traced to a Chinese content farm began to spread on Taiwan's social media platforms that the Tsai administration, out of concern for the environment, planned to ban firecrackers and the burning of traditional "ghost money" and incense.[62] The rumor sparked some 10,000 people to take to the streets of Taipei protesting this violation of traditional Taoist, Buddhist, and traditional cultural practices.[63] In the 2018 Taiwan election, Chinese disinformation tended to concentrate on the political center of Taiwan politics and tacitly encourage political fringes.[64]

Russia has, perhaps, an even more established track record of using social media to inflame tensions and generally cause chaos. Russian agents have promoted wide-ranging causes, from Texas secessionism to Bosnian Serb nationalism, to "effectively aggravate the conflict between minorities and the rest of the population."[65] During the early stages of the Ukraine crisis, Russian Military Intelligence (GRU)

---

[60] Shin Bo-Young [신보영], "North Korea Appropriates South Korean RRN for Coordinated Propaganda [북, 남한 주민번호 도용 네티즌 조직적 선동]," *Culture Daily* [문화일보], 2010.

[61] Interviews with Chinese disinformation researchers, Taipei, Taiwan, January 2019.

[62] Cole, 2017; Lu et al., 2017.

[63] "Taiwan's Taoists Protest Against Curbs on Incense and Firecrackers," BBC News, July 23, 2017.

[64] Interview a Chinese disinformation expert, Taipei, Taiwan, January 2019.

[65] *United States of America v. Elena Alekseevna Khusyaynova, No. 1:18- MJ-464,* U.S. District Court, Alexandria, Va., September 28, 2018, p. 13; Tim Lister and Clare Sebastian, "Stoking Islamophobia and Secession in Texas—From an Office in Russia," CNN, October 6, 2017; David Salvo and Stephanie De Leon, "Russia's Efforts to Destabilize Bosnia and Herzegovina," Alliance for Securing Democracy, April 25, 2018.

psychological operations officers attempted to galvanize pro-Russian Ukrainians by disseminating messages on social media that claimed "brigades" of "zapadentsy" (Westerners) were going to "rob and kill" other Ukrainians, adding that the protestors attempting to unseat Viktor Yanukovych, an ally of the Kremlin, were "completely different" from ordinary Ukrainians.[66] More recently, in September 2018, Russia used social media to spread fake information (based on the fictitious murder of a Ukrainian boy by Hungarians, according to Ukrainian experts) about ethnic clashes between Ukrainians and Hungarians in Western and Central Ukrainian sites.[67]

North Korea has tried to use disinformation to drive wedges in South Korean society. Beginning around 2010 and culminating in 2013, the popular South Korean internet platform dcinside—on which users can post to bulletin boards organized by topic—detected a series of suspicious posts using various proxy Internet Protocols and posting aliases, peaking at around 900 posts per day.[68] These posts were similarly formatted, with news and blog pieces taken from elsewhere interlaced with commentary on a variety of topics of interest in North Korea, including criticizing President Park Geun-hye, the U.S. military presence in South Korea, and United Nations–backed sanctions of North Korea.[69] South Korean officials were never able to definitively link these actions back to specific entities in North Korea, but the originators of the posts presumably intended to drive a rift in the U.S.–South Korea alliance and inflame internal South Korean political tensions.

In all these cases, disinformation can only go so far. For disinformation to be credible, it must be built on reality at some level. Russian intimidation of Ukraine, Chinese intimidation of Taiwan, and (to a lesser extent) North Korea's attempt to intimidate South Korea

---

[66] Ellen Nakashima, "Inside a Russian Disinformation Campaign in Ukraine in 2014," *Washington Post*, December 25, 2017.

[67] Interview with security services, Kyiv, Ukraine, March 6, 2019.

[68] Kim Jung-Woo [김정우], "North Korea's 'Internet Invasion' Is Flaring Up [최근 기승 부리는 북한의 '인터넷 남침]," *Chosun* [월간조선], 2013.

[69] Kim Jung-Woo, 2013; dcinside management, "Inquiry Regarding North Korea's Disinformation Campaign," email correspondence with authors, July 9, 2019.

would not have the same resonance but for the fact that all three countries have the military capacity to hurt their neighbors. Similarly, their efforts to discredit the Ukrainian, Taiwan, or South Korean governments or exacerbate social cleavages would not work unless these efforts had some basis in reality. Russia has shifted its disinformation tactics away from generating new content to amplifying real, albeit fringe, opinions—the latter is not only harder to detect as foreign disinformation campaign but also builds on a preexisting basis of support.[70]

It is perhaps unsurprising, then, that there are fewer examples of disinformation campaigns successfully persuading hostile populations to abandon their former ideas and adopt new ones. Russia certainly tried to promote Russian-Ukrainian "brotherhood" both before and after the conflict, but these efforts have been largely unsuccessful, especially when other Russian actions belie that narrative.[71] Arguably, the same has been true of Chinese efforts to persuade Taiwan to accept unification with China and of North Korea's interest in persuading South Korea to reunify on the North's terms. Bluffing only gets one so far—especially when the facts make the bluff obvious.

Not All Social Media Platforms Are Created Equal

Even in a globally interconnected world, the social media environment still takes on regional and local characteristics; so does the disinformation fight. Given the mismatch between DoD's global scope and its limited resources, there might be a temptation to concentrate resources on a handful of the most-popular platforms worldwide—Facebook, Twitter, and the like—and with good reason. After all, as depicted in Figure 2.2, Facebook dwarfs other platforms in terms of its total user base and global reach. In some places, such as the Philippines, Facebook is synonymous with the internet itself. And yet, it is a mistake to equate the size of a user base to relevancy in the disinformation fight.

---

70 Interview with technology analyst, Washington, D.C., February 5, 2019; interview with academic, Washington, D.C., February 11, 2019.

71 Interview with security officials, Kyiv, Ukraine, March 6, 2019.

Figure 2.2
Social Media Platform Users over Time

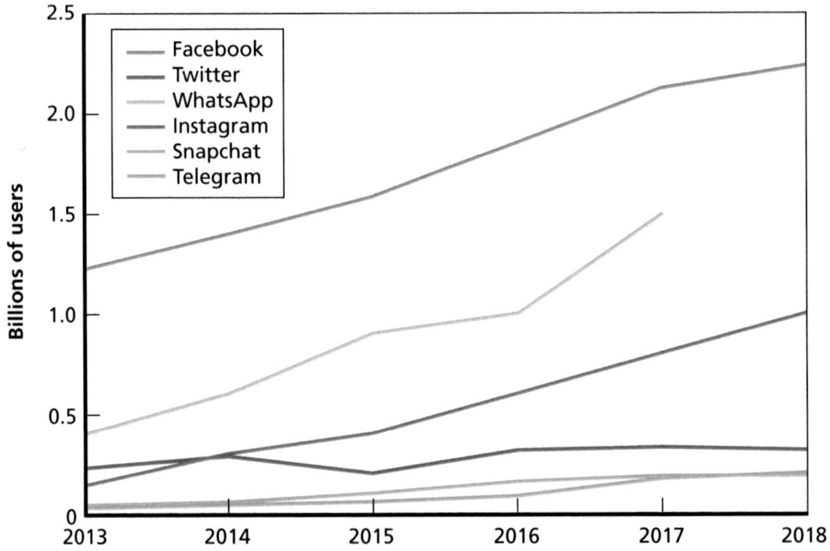

SOURCES: Statista, "Number of Monthly Active Facebook Users Worldwide as of 3rd Quarter 2018 (in millions)," webpage, October 2018g; Statista, "Number of Monthly Active Twitter Users Worldwide from 1st Quarter 2010 to 4th Quarter 2018 (in millions)," webpage, February 2019; Statista, "Number of Monthly Active WhatsApp Users Worldwide from April 2013 to December 2017 (in millions)," webpage, January 2018a; Statista, "Number of Monthly Active Instagram Users from January 2013 to June 2018 (in millions)," webpage, June 2018e; Statista, "Number of Daily Active Snapchat Users from 1st Quarter 2014 to 3rd Quarter 2018 (in millions)," webpage, October 2018f; Statista, "Number of Monthly Active Telegram Users Worldwide from March 2014 to March 2018 (in millions)," webpage, March 2018c.

Although U.S. social media platforms might be the most popular globally, the United States does not have a monopoly on social media companies. Russians have Odnoklassniki and VKontakte (VK). China has WeChat and Weibo. Russian and Chinese social media companies are also suspected of having ties with the security services of their countries; if this is true, these companies inherently present greater risks regarding disinformation campaigns.[72] Some of these platforms are popular outside Russia and China, including in regional hotspots. For example, VK was the most popular social media site in Ukraine until the Ukrainian government banned the site in May 2017—a ban largely resulting from the site's close connections to Russian security.[73] Even after the ban had been in effect for a year, VK remained the fourth most popular site in Ukraine.[74]

Identifying the key terrain in the social media disinformation fight is more complex though than simply looking for Russian- or Chinese-owned platforms. Some direct messaging platforms—such as LINE or WhatsApp, with its end-to-end encryption—are inherently hard to monitor for disinformation. In some parts of the world, such as South Asia, large-scale WhatsApp groups are the primary means by which the public gets news.[75] This can make these platforms particularly attractive vectors for disinformation. For example, one expert from Taiwan argued that LINE is a relatively fertile ground for PLA disinformation operations because messages can circulate for some time before the government becomes aware of them and responds.[76] Taiwan media suggested that part of the PLA's political interference in the November 2018 election occurred on LINE.[77]

Other platforms—such as PTT in Taiwan or dcinside in South Korea—attract relatively small audiences globally but are key battlegrounds for the information fight inside those countries because they

---

[72] Jennifer Monaghan, "Vkontakte Founder Says Sold Shares Due to FSB Pressure," *Moscow Times*, April 17, 2014.

[73] "Ukraine Bans Its Top Social Networks Because They Are Russian," *The Economist*, May 19, 2017.

[74] Interfax-Ukraine, "Banned VK Social Network 4th in Internet Traffic in Ukraine in April," *Kyiv Post*, May 17, 2018.

[75] See, for example, Timothy McLaughlin, "Disinformation Is Spreading on WhatsApp in India—And It's Getting Dangerous," *The Atlantic*, September 5, 2018.

[76] Interview with multiple Taiwan government officials and Chinese disinformation researchers, Taipei, Taiwan, January 2019.

[77] Chung Li-hua and William Hetherington, "China Targets Polls with Fake Accounts," *Taipei Times*, November 5, 2018; Keoni Everington, "China's 'Troll Factory' Targeting Taiwan with Disinformation Prior to Election," *Taiwan News*, November 5, 2018.

reach certain key audiences. For example, interviewees from Taiwan commented that Beijing uses PTT as the preferred vector to reach those under 40 years of age; other interviewees saw Facebook as the primary vector to reach younger audiences while LINE was more likely to be used to communicate with older social media users in Taiwan.[78] dcinside operates similarly in North Korea. The platform averages around 700,000 to 900,000 posts per day on specific topics (which allows North Korea to target a subset of the population), and it allows users to post on the platform anonymously without a sign-up process or, as in 2012, requiring South Korean users to provide their RRNs.[79]

Smaller platforms also might lack the resources or interest in fighting disinformation that larger companies have. 4chan, for example, allows anonymous posting and has cultivated a culture of unfiltered content, making the site attractive to fringe groups and conspiracy theorists.[80] Reddit, similarly, is considerably smaller than Facebook and has fewer resources to devote to counter disinformation; consequently, it has struggled to mount timely responses.[81] This all means that disinformation can start on a smaller platform—such as reddit or 4chan—and then jump over to larger platforms. By the time disinformation makes the leap into the mainstream, it could be too late to mount an effective response.[82]

Russia's interference in the 2017 French election provides a good example of this pattern. Russia allegedly participated in the hack of Emmanuel Macron's campaign and subsequently leaked the information to Archive.org, PasteBin, and 4chan.[83] From there, the informa-

---

78 Interview with China disinformation expert, Taipei, Taiwan, January 2019.

79 dcinside management, 2019.

80 Interview with academic, Washington, D.C., February 22, 2019; also see Emma Grey Ellis, "4chan Is Turning 15—and Remains the Internet's Teenager," *Wired*, June 1, 2018.

81 Interview with think tank analyst, Washington, D.C., February 26, 2019; Ben Collins, "On Reddit, Russian Propagandists Try New Tricks," NBC News, September 25, 2018; Craig Silverman and Jane Lytvynenko "Reddit Has Become a Battleground of Alleged Chinese Trolls," *BuzzFeed News*, March 14, 2019.

82 Interview with think tank analyst, Washington, D.C., February 26, 2019.

83 Vilmer et al., 2018, pp. 106–116.

tion was picked up and spread by Twitter.[84] Other false accusations against Macron followed a similar pattern. The #MacronGate and #MacronCacheCash allegations that stated Macron had a secret offshore account first surfaced on 4chan and spread through Twitter later.[85]

Ultimately, although disinformation on social media might be a global phenomenon, the primary battle ground varies by region. Depending on where the U.S. military is operating, it will need to identify key terrain in this virtual battlefield and recognize that terrain might not simply be Facebook and Twitter.

Disinformation's Effectiveness Remains Unknown

Disinformation campaigns on social media are commonly likened to having the same effect on 21st-century warfare that a variety of kinetic weapons had on 20th-century conflicts. A podcast produced by the Wharton School of University of Pennsylvania, for example, argues that "social media is the new weapon in modern warfare."[86] Former Senate aide Mike Ongstand goes a step further, arguing that disinformation on social media has the same game-changing effects on modern warfare as the machine gun had during World War I.[87] A report produced for the Department of State's GEC refers, somewhat cutely, to disinformation efforts as "weapons of mass distraction."[88] Multiple Taiwan officials interviewed for this report claimed disinformation campaigns on social media are "as powerful as a missile" or

---

[84] Vilmer et al., 2018, pp. 106–116.

[85] Vilmer et al., 2018, pp. 106–116.

[86] Knowledge@Wharton, "Why Social Media Is the New Weapon in Modern Warfare," January 17, 2019.

[87] Mike Ongstad, "Social Media Is the Machine Gun of Modern Disinformation War," *The Hill*, October 26, 2018.

[88] Christina Nemr and William Gangware, *Weapons Of Mass Distraction: Foreign State-Sponsored Disinformation in the Digital Age*, Washington, D.C.: Park Advisors, March 2019.

even more so.[89] For all analogies out there, however, disinformation on social media fundamentally differs from machine guns and weapons of mass destruction in a variety of ways—perhaps most notably in that we still do not know how effective it actually is despite all the attention this topic has received over the past several years. The effect of kinetic weapons are governed by the laws of physics, so we can quantify the relative impact of a machine gun or missile. This is not the case for disinformation, which is governed by the far more nebulous domain of human psychology.

Many of the more-detailed analyses about the impact of these campaigns simply punt on the question of effectiveness. For example, the Director of National Intelligence concluded that Russia launched a large-scale effort to sway the 2016 elections but avoided answering whether these efforts actually made any difference, saying that the intelligence community "does not analyze US political processes or US public opinion."[90] The director at the time, Daniel Coats, did state that "there should be no doubt that Russia perceives its past efforts as successful."[91] A similarly detailed study by the House of Commons in the United Kingdom also clearly identified Russian disinformation attempts to sway British elections and the Brexit referendum but could not reach a definitive judgement on the results of this effort. Instead, the study argues, "It is surely a sufficient matter of concern that the Government has acknowledged that interference has occurred, irrespective of the lack of evidence of impact."[92]

There are examples of disinformation producing narrow operational effects. For example, during the height of the Ukraine con-

---

89 Interview with a Taiwan government official, Taipei, Taiwan, January 2019; interview with a Taiwan academic, Taipei, Taiwan, January 2019.

90 Office of the Director of National Intelligence, "Background to *Assessing Russian Activities and Intentions in Recent US Elections*: The Analytic Process and Cyber Incident Attribution," Washington, D.C., January 6, 2017b.

91 Matthew Rosenberg, Charlie Savage, and Michael Wines, "Russia Sees Midterm Elections as Chance to Sow Fresh Discord, Intelligence Chiefs Warn," *New York Times*, February 13, 2018.

92 House of Commons Digital, Culture, Media, and Sport Committee, *Disinformation and 'Fake News': Final Report*, London, February 18, 2019, p. 70.

flict, Russia used disinformation to spur calls to Ukrainian soldiers via cellphones, helping Russian forces geolocate Ukrainian troop formations and spurring Russian lethal targeting. As Colonel Liam Collins explains in his analysis of Russian operations:

> [Ukrainian] soldiers receive texts telling them they are "surrounded and abandoned." Minutes later, their families receive a text stating, "Your son is killed in action," which often prompts a call or text to the soldiers. Minutes later, soldiers receive another message telling them to "retreat and live," followed by an artillery strike on the location where a large group of cellphones was detected.[93]

In these cases, disinformation became the bait for lethal action, and what started as fake news became a tragic reality. For their part, Ukrainians claim their own disinformation campaigns prompted the defection of a separatist commander in the Donbass and the removal of a Russian separatist leader from command.[94] These examples notwithstanding, the war in Eastern Ukraine was not won—or lost—based on the success of disinformation; the outcomes stemmed from the conventional balance of forces in the Donbass.[95]

---

93 Liam Collins, "Russia Gives Lessons in Electronic Warfare," Association of the United States Army, July 26, 2018.

94 Interview with a politician, Kyiv, Ukraine, March 5, 2019; interview with security officials, Kyiv, Ukraine, March 6, 2019. For a similar account about the misdeeds and later mysterious deaths of senior leaders in the Donbass (albeit without the Ukrainian information operation included), see Jack Losh, "Is Russia Killing Off Eastern Ukraine's Warlords?" *Foreign Policy*, October 25, 2016.

95 A detailed RAND Corporation military analysis of Crimea and Eastern Ukraine concluded, "Unfortunately, it is difficult to discern any tangible operational advantages Russia gained from its information campaign during the Crimean annexation." In Eastern Ukraine, the strategic effects were even more ambiguous. "Russia's information war in Eastern Ukraine polarized the population, but ultimately Ukraine proved infertile ground for separatism . . . ." Consequently, "Moscow devoted an increasing amount of resources to the conflict, ultimately escalating it to a conventional war with its own regular units in the lead." See Michael Kofman, Katya Migacheva, Brian Nichiporuk, Andrew Radin, Olesya Tkacheva, and Jenny Oberholtzer, *Lessons from Russia's Operations in Crimea and Eastern Ukraine*, Santa Monica, Calif.: RAND Corporation, RR-1498-A, 2017, pp. 29, 76.

Chinese disinformation campaigns in Taiwan have similarly measurable operational effects. As mentioned earlier in this chapter, Chinese disinformation prompted a protest about false rumors of Taiwan restricting cultural and religious practices and indirectly led to a Taiwan consul committing suicide after being falsely accused of mishandling the response to a natural disaster.[96] The most quantifiable aspect of disinformation's impact, perhaps, was best described by a senior Taiwan official, who said disinformation ate up senior policymakers' limited time by forcing them to respond to every false or misleading story.[97]

For all these apparent operational-level victories, it is hard to show that Russian disinformation in Ukraine or Chinese disinformation in Taiwan produced strategic-level effects. We can document Russian attempts to undermine Ukrainian morale, hinder mobilization, and generally impede military effectiveness during the height of the fighting in the Donbass in 2014 and 2015, but it difficult to parse the military effect of these efforts from the real chaos in the Ukrainian high command (which had just experienced the Maidan revolution). We do know that more-recent Russian attempts to emphasize a common Russian-Ukrainian "brotherhood" and blame a small pro-Western faction for conflict have largely fallen on deaf ears.[98] Similarly, despite decades of sustained Chinese and North Korean information warfare—first through traditional means and now through social media—Taiwan's public has yet to support unifying with the mainland and the South Korean public does not support a North Korea–led reunification.

Given the multitude of factors that influence human decision-making, the impact of any psychological operation will, by its very nature, have more-ambiguous effects than a machine gun or a nuclear weapon. If nothing else, however, the prevalence and scale of some of these operations make it likely that at least some of these efforts succeeded in changing the opinions of some among their targeted audience or prompting them to take actions they might not have otherwise. At the same time, the fact that we are still struggling to prove the

96 "Taiwan's Taoists Protest . . . ," 2017; Ko Tin-yau, 2018.

97 Interview with senior Taiwan government official, Taipei, Taiwan, January 2019.

98 Interview with security officials, Kyiv, Ukraine, March 6, 2019.

impact of these efforts after all the research that has been done on this subject at least raises the question of just how powerful disinformation on social media is as a weapon—to the extent it should be considered a weapon at all.

Clearly, disinformation on social media can produce operational effects—as demonstrated in Russia's use of disinformation to aid targeting in Ukraine or China's use of disinformation to provoke protests in Taiwan. But for now, at least, there is no clear evidence that these campaigns can shift popular opinions to the point that consumers acquire views diametrically opposed to their strongly held prior assumptions: Chinese disinformation cannot persuade Taiwan to rejoin the mainland; South Koreans are still unwilling to reunite with the North under a North Korean flag. What remains unknown is the extent to which disinformation on social media can shift opinion regarding issues on which the target audience does not have strongly held prior beliefs, or whether social media can sway audiences to ideas that they might be predisposed to believe.

Ultimately, how we view disinformation's effectiveness might depend on our baselines. If we view disinformation on social media as a magic panacea of sorts, then it is sure to disappoint. But if we base our evaluation of effectiveness on what would occur absent any sort of disinformation campaign, it is sure to exceed expectations.

What's Next: More of the Same

Despite the ambiguous success rate of disinformation campaigns on social media, there is near universal consensus among the experts we interviewed around the globe that these types of campaigns will continue and possibly even increase. There are plenty of reasons to believe that disinformation will become a more prominent tool of state power in the coming years.

First, these types of campaigns are relatively cheap. While verifiable budget numbers are difficult to come by, some place China's budget for propaganda at $10 billion, with much of that for domestic consumption.[99] Even if the full $10 billion went to external influence campaigns on social media, the sum would pale in comparison with the overall size of the China's defense budget—officially valued at $177.5 billion in 2019 and likely considerably higher.[100]

Russia's spending on foreign influence is similarly opaque. and what data are available tend to be partial and ad hoc. One 2017 survey by a Russian cybersecurity firm with ties to the defense ministry claimed that Russia's overall budget for cyberoperations, which include "information wars" and cyberattacks designed to affect the "mood and behavior" of civilian populations, amounted to around $300 million annually.[101] Estimates for Russian spending on more overt forums, such as media outlets RT and Sputnik, range from $190 million to $500 million.[102] Oxford University's Computational Propaganda Project's analysis suggests that the IRA spent a mere $74,000 on advertisements to influence target audiences on Facebook.[103] Even if the actual numbers are higher than the publicly available ones, disinformation campaigns on social media remain orders of magnitude cheaper than many other tools of power, possibly making it worth the gamble even if the return on investment remains ambiguous.

Second, disinformation campaigns on social media require relatively little infrastructure. While more-sophisticated disinformation campaigns are tailored to the individual and require the ability to understand the linguistic and cultural nuances of an audience, lower-grade disinformation does not require the same access to intelligence and in-depth knowledge. Perhaps the best example of the relatively low

---

[99] Shambaugh, 2015.

[100] China Power, "What Does China Really Spend on Its Military?" Center for Strategic and International Studies, undated.

[101] Viktoria Nosova, "Study: Russia Is Among the Top Five Countries with the Strongest Cyber Force," Vesti.ru, October 10, 2017.

[102] Warren Strobel, "U.S. Losing 'Information War' to Russia, Other Rivals: Study," Reuters, March 25, 2015.

[103] Philip N. Howard, Bharath Ganesh, and Dimitra Liotsiou, "The IRA, Social Media and Political Polarization in the United States, 2012–2018," Oxford Internet Institute, December 17, 2018, Table 4.

barriers to entry for disinformation come from the Philippines, where former offshore gaming companies and business call centers have been converted into troll farms or "click factories."[104] These for-hire operations are staffed not by particularly savvy information operators but by relatively low-paid workers who try to boost the profile for businesses and Filipino domestic politicians by operating fake Facebook or Twitter accounts.[105]

Disinformation campaigns do not even need a formal infrastructure. Russian disinformation campaigns, for example, are conducted by a wide swath of unaffiliated actors, including patriotic groups that ideologically align with some or all of Russia's policy objectives.[106] With relative ease, private individuals searching the internet in Russian can find a black market for fake Facebook accounts—including some that have been groomed by individuals posting genuine material for years to give these accounts seeming credibility.[107] Potential buyers can purchase accounts wholesale for their own purposes or "buy likes"—that is, have these fake accounts promote selected material—to falsely inflate a post's popularity.[108] In other words, there are relatively few barriers to entry that would prevent any actor from launching a disinformation campaign.

Third, from a societal perspective, U.S. vulnerability—like that of many other countries around the world—might be increasing. Recent RAND research suggests that the United States is suffering from "Truth Decay," as the line between fact and opinion becomes increasingly blurred. There is declining faith in formerly trusted

---

[104] Interview with political and defense analysts, Manila, Philippines, May 18–19, 2019.

[105] Jonathan Head, "Outlaw or Ignore? How Asia Is Fighting 'Fake News,'" BBC, April 4, 2018; interview with political and defense analyst, Manila, Philippines, May 2019.

[106] Elizabeth Bodine-Baron, Todd C. Helmus, Andrew Radin, and Elina Treyger, *Countering Russian Social Media Influence*, Santa Monica, Calif.: RAND Corporation, RR-2740-RC, 2018, p. 10.

[107] Interview with law enforcement, Kyiv, Ukraine, March 7, 2019; interview with data analytics firm, Kyiv, Ukraine, March 9, 2019.

[108] Interview with law enforcement, Kyiv, Ukraine, March 7, 2019; interview with data analytics firm, Kyiv, Ukraine, March 9, 2019.

sources of truth. In some cases, people fundamentally disagree over the facts themselves. Arguably, these trends contribute to an environment conducive to disinformation.[109]

Finally, technological trends suggest that disinformation will become harder to detect in the coming years. The proliferation of artificial intelligence, machine learning, and "deep fakes"—falsified videos and audio that look and sound like legitimate material—will open new frontiers in disinformation, making these campaigns an increasingly attractive option for would-be adversaries looking to sow chaos and division.[110] Should these techniques be perfected faster than the means to detect them, disinformation could become a significantly greater problem than it is now.

In sum, the United States and its allies and partners will be confronting disinformation campaigns on social media for some time to come. Whether the United States—or any other state for matter—are prepared for these challenges, remains an unanswered question, as we will discuss in the next chapter.

---

[109] Jennifer Kavanagh and Michael D. Rich, *Truth Decay: An Initial Exploration of the Diminishing Role of Facts and Analysis in American Public Life*, Santa Monica, Calif.: RAND Corporation, RR-2314-RC, 2018.

[110] Robert Chesney and Danielle Citron, "Deepfakes and the New Disinformation War: The Coming Age of Post-Truth Geopolitics," *Foreign Affairs*, January/February 2019.

CHAPTER THREE

A Divided and Uncertain Response

Since the 2016 election interference, the United States has been locked in an information war with Russia. At least, that is how late Senator and former Chairman of the Senate Armed Services Committee John McCain, former United Nations Ambassador Nikki Haley, Vice President Richard Cheney, and some congressional Democrats have described it.[1] And yet, three years later, it is still unclear who inside USG should lead this effort, how this war should be fought, or whether the term "war" even applies. Different efforts to combat disinformation on social media are being made by the USG, U.S. allies and partners, and industry and the private sector. Some of these efforts focus on increasing societal resilience and making the population savvier consumers of what they read and see online; others aim to decrease the power of the disinformation—improving efforts to detect, debunk, and curb the spread of campaigns. Still other efforts target the source the disinformation—threatening sanctions, prosecution, and various retaliatory measures. At best, these measures are loosely coordinated and complementary; at worst, they work at cross purposes.[2] Speaking

[1] Theodore Schleifer and Deirdre Walsh, "McCain: Russian Cyberintrusions an 'Act of War,'" CNN, December 30, 2016; Alex Lockie, "Dick Cheney: Russia Meddling in the US Election Could Be 'an Act of War,'" *Business Insider*, March 28, 2017; John Haltiwanger, "Russia Committed Act of War With Election Interference, Nikki Haley Says," Newsweek, October 19, 2017; Julia Manchester, "Dem Calls Russia Meddling 'Act of War,' Urges Cyber Attack on Moscow Banks," *The Hill*, July 17, 2018.

[2] For a similar conclusion about the fragmentation of the U.S. response to Russian disinformation in particular, see Bodine-Baron et al., 2018.

to experts across the world from many different backgrounds, there is little consensus about the way ahead in the counterinformation fight, except in one respect: There is no magic panacea for disinformation, nor is there likely to be one any time soon.

Categorizing the Responses

In theory, there are several ways to try to stop a disinformation campaign. One way is to deter nefarious actors from producing disinformation either because they fear the consequences of engaging in such behavior (*deterrence by punishment*) or because they do not believe a disinformation campaign would succeed (*deterrence by denial*).[3] Another method is to stop or limit the spread of the disinformation by removing the content or burying it so that it does not show up in search results. A third option is to prevent the content from resonating by proactively inoculating the target audience against untruthful information or by reactively exposing the disinformation as a ploy. In sum, as disinformation follows its life cycle, actors must adopt different countermeasures to mitigate it (see Box 3.1).

Production countermeasures are those that aim to prevent actors from producing or ordering the production of disinformation by threatening economic sanctions, diplomatic isolation, or criminal indictments so that costs of the action exceed the gains. Production countermeasures can also involve making actors believe they will fail to attain their goals by making disinformation operations on social media less profitable or more difficult to conduct—for example, by launching offensive cyberattacks to block troll farms.

---

[3] Glenn Herald Snyder, *Deterrence and Defense: Toward a Theory of National Security*, Princeton, N.J.: Princeton University Press, 1961. The scholarly literature distinguishes between *deterrence by denial*, which refers to measures taken prior to an attack, and *defense*, which refers to measures taken once an attack is occurring. However, as prior RAND work points out, this distinction is not very helpful with regard to activities that tend to be continuous rather than discrete incidents. (See Bodine-Baron et al., 2018, p. 21.) In our study, we treated possible defense measures as deterrence by denial.

Box 3.1
Countermeasures to Disinformation

Production [prevent actors from producing or ordering production of content]	**Distribution** [restrict actors from distributing content]	**Consumption** [build audience resilience, lower susceptibility to content]
• Deterrence by denial (e.g., cyberattacks on troll farms) • Deterrence by punishment (e.g., threats of prosecution, sanctions, and other retaliation)	• Blocking of actors • Banning or restricting social media networks channels • Algorithmic, legal, and manual limits on spread of disinformation	• Debunking • Media literacy • Proactive public diplomacy • Positive strategic communication and message discipline • Reducing credibility of messengers and messages
Detection or Awareness-Raising • Identifying and analyzing the actors and mechanisms inside the disinformation life cycle • Raising awareness of threat among decisionmakers and other audiences		
Institution-Building • Creating institutions with authorities and capabilities to combat disinformation		

Responses can also target the *distribution* of disinformation. Perhaps more than any other category of responses, distribution countermeasures often fall to the private sector—the social media companies themselves—because these efforts include blocking or downgrading disinformation in search results, shutting down fake accounts, and (in their most extreme form) banning social media use and disinformation-prone social media platforms entirely.

Consumption countermeasures can come in two forms. Some try to proactively build societal resilience through positive public diplomacy efforts (that create preexisting favorable attitudes toward a given subject or institution) and media literacy programs (that train audiences to spot disinformation and generally be wary of what they read on social media). Consumption countermeasures can also be more reactive, debunking or exposing disinformation to reduce the credibility of the messages or the messengers.

Beyond the production-to-consumption chain, there are at least two prerequisites for any successful countermeasure. First, there is *detection and awareness-raising*. As with any threat, a successful policy response requires an in-depth understanding of how disinformation campaigns work, who are the actors behind them, and what they hope to accomplish. These measures involve identifying and analyzing the actors and mechanisms throughout the disinformation life cycle. Second, there is *institution-building*. All the efforts discussed so far require organizations and structures with the requisite authorities and capabilities to conduct any of the actions described. Therefore, part of the response to countering disinformation campaigns involves building new institutions, recruiting the right people, drafting the right policies, and providing the right tools to make these efforts work.

As we shall see, public and private actors have placed different emphases on each of these five categories. Nation-states—including the United States—have a hand in all five categories of countermeasures and are among the few actors that can target the production phases. The social media companies likely play the most important role in the detection and distribution phases. Finally, civil society groups—investigative journalists, advocacy groups, and think tanks—often concentrate on the detection-and-awareness and consumption phases.

U.S. Government Efforts Lack Coherence and Coordination

Who in the United States government is responsible for combating the foreign influence on social media? There is no straightforward answer. Since the end of the Cold War, the United States lost a single entity for combating foreign information campaigns, dividing the responsibility instead through multiple different agencies each with partial responsibilities for counter-disinformation and each targeting different stages of disinformation life cycle (see Box 3.2) but with no single USG agency tying these efforts together.

During the Cold War, most U.S. efforts in the information environment were carried out by a single executive branch organization, the U.S. Information Agency (USIA). Established in 1953 the wake of World War II, the USIA "played a bellwether role in developing and carrying out a national strategy for overseas information and cultural operations."[4] As part of these operations, USIA operated the Voice of America (VOA) broadcasting network; produced foreign-language literature, such as magazines and leaflets, and other media, such as films; and conducted cultural-exchange and English-language programs.[5]

With the collapse of the Soviet Union and end of the Cold War, the United States no longer needed to counter the propagation of communist ideology and no longer believed that it needed an agency for information.[6] It dissolved the USIA and divided its functions between the Department of State and a new entity—the Broadcasting Board of Governors.[7] The former took over the public diplomacy and edu-

---

[4] Wilson P. Dizard, Jr., *Inventing Public Diplomacy: The Story of the U.S. Information Agency*, London: Lynne Rienner Publishers, 2004, p. 4.

[5] Dizard, 2004, pp. 4–5.

[6] Yahya R. Kamalipour and Nancy Snow, eds., *War, Media, and Propaganda: A Global Perspective*, Lanham, Md.: Rowman & Littlefield, 2004, p. 221.

[7] Matthew Armstrong, "No, We Do Not Need to Revive the U.S. Information Agency," War on the Rocks, November 12, 2015.

Box 3.2
Who in the USG Does What

Production [prevent actors from producing or ordering production of content]	**Distribution** [restrict actors from distributing content]	**Consumption** [build audience resilience, lower susceptibility to content]
• Diplomatic pressure/sanctions by Departments of Treasury and State • Indictments by Department of Justice	• Pressure on social media companies by USG to crack down on spread of disinformation	• Proactive messaging to foreign audiences from U.S. Agency for Global Media (USAGM) and the Department of State's public diplomacy wing • Countering false information about disasters by Department of Homeland Security's Social Media Working Group
Detection and Awareness-Raising		
• Detection of foreign disinformation by the intelligence community and Federal Bureau of Investigation (FBI)'s foreign influence task force		
Institution-Building		
• Multiple agencies created new organizations and task forces to address this problem, most notably Department of State's GEC		

NOTE: DoD actions are covered in Table 3.3 and so are not depicted here.

cational and cultural programs.[8] The Broadcasting Board of Governors (or, as of 2018, the USAGM), inherited the USIA's international broadcasting functions and with it, the authority to oversee the USG's existing broadcasting services, such VOA, Radio Asia, and Radio Free Europe/Radio Liberty. In other words, after the demise of the USIA, U.S. capability to proactively shape foreign audiences' perceptions and reduce their susceptibility to any hostile disinformation was divided between two separate bureaucracies.[9]

In March 2016, the Department of State established yet another new entity, the GEC,[10] which was originally designed to counterterrorist-sponsored messaging. Spurred by the threat of Russian information campaigns, Congress expanded the GEC's mandate to include countering state information campaigns.[11] Specifically, Congress instructed that the GEC "recognize, understand, expose, and counter foreign state and foreign nonstate propaganda and disinformation efforts aimed at undermining or influencing the policies, security, or stability of the United States and the United States allies and partner nations."[12] Broadly speaking, the GEC has three missions today:

1. identifying and analyzing foreign propaganda and disinformation
2. planning and executing efforts to respond to or counter propaganda and disinformation

---

[8] For additional details about the handoff of nonbroadcasting USIA activities to the Department of State, see U.S. Advisory Commission on Public Diplomacy, *Consolidation of USIA into the State Department: An Assessment After One Year*, Washington, D.C.: U.S. Department of State, October 2000. Also see Armstrong, 2015.

[9] USAGM, "History," webpage, undated; USAGM, *FY 2018 Performance and Accountability Report*, Washington, D.C.: U.S. Broadcasting Board of Governors, November 2018, p. 12.

[10] Obama, Barack, "Executive Order—Developing an Integrated Global Engagement Center to Support Government-Wide Counterterrorism Communications Activities Directed Abroad and Revoking Executive Order 13548," Washington, D.C.: White House, Executive Order 13721, March 14, 2016.

[11] 22 U.S.C. § 2656.

[12] 22 U.S.C. § 2656.

3. evaluating the effectiveness of the GEC's activities and identifying shortcomings in U.S. capabilities to counter and respond to foreign propaganda efforts.[13]

Other government agencies also have roles in the counter-disinformation effort. The FBI serves as the lead USG agency responsible for investigating foreign influence operations, defined as "covert actions by foreign governments to influence U.S. political sentiment or public discourse."[14] While such efforts have taken many forms, the most recent ones predominantly involve adversaries' use of "false personas and fabricated stories on social media platforms to discredit U.S. individuals and institutions."[15] In 2017, the FBI launched the Foreign Influence Task Force to identify and counter all foreign influence operations aimed at the United States, including adversary-sponsored disinformation efforts. The task force has established a webpage to inform the U.S. public about disinformation campaigns and cyberattacks.[16]

Additionally, the Department of Homeland Security's Social Media Working Group for Emergency Services and Disaster Management is responsible for establishing best practices for countering attempts to deliberately propagate false information via social media about not only emergencies and disasters but also about response and recovery efforts.[17]

Aside from these defensive measures, other parts of the USG have tried to punish those responsible for creating and disseminating disinformation as a way to deter future campaigns. For example, the Department of Treasury's Office of Foreign Assets Control targeted the Russian intelligence operatives believed to be behind the 2016 elec-

---

[13] 22 U.S.C. § 2656.

[14] FBI, "The FBI Launches a Combating Foreign Influence Webpage," August 30, 2018.

[15] FBI, 2018.

[16] FBI, 2018.

[17] Social Media Working Group for Emergency Services and Disaster Management, *Countering False Information on Social Media in Disasters and Emergencies*, Washington, D.C.: Department of Homeland Security, March 2018; Science and Technology Directorate, "Social Media Working Group for Emergency Services and Disaster Management," fact sheet, Washington, D.C.: U.S. Department of Homeland Security, November 22, 2017.

tion interference with sanctions, while the Department of Justice has pursued criminal indictments against these individuals.[18]

Lastly, assuming USG information efforts reflect those of the Cold War environment, it is likely that U.S. intelligence agencies play a role in U.S. information efforts. For much of the Cold War era, the Central Intelligence Agency monitored Soviet disinformation efforts.[19] In 1981, the USG established the Active Measures Working Group, an interagency body responsible for coordinating the activities of different USG actors conducting information efforts.[20] The unclassified 2017 Intelligence Community Assessment indicates that, at the very least, U.S. intelligence agencies collect and analyze intelligence related to foreign use of information efforts to target the United States.[21]

Ultimately, the real question is whether all these varied efforts across the interagency are effective. Certainly, the overarching statistics are impressive. As of December 2018, the GEC (theoretically the linchpin of the USG's counter-disinformation efforts) had "25 initiatives in 21 countries designed to counter Russian propaganda efforts," which involved such activities as "supporting independent local news and civil society organizations with everything from propaganda-sensitivity training to data analysis exposing Russian subversion."[22] The GEC also has teams devoted to countering malign information efforts perpetrated by China, Iran, and terrorist organizations.[23] In 2018, the GEC unveiled the Information Access Fund to provide grants to nongovernmental organizations (NGOs), research centers, scholars, media organizations,

---

[18] Nathan Layne, "U.S. Imposes Fresh Russia Sanctions for Election Meddling," Reuters, December 19, 2018; *United States of America v. Internet Research Agency et al.*, U.S. District Court, District of Columbia, February 16, 2018.

[19] Theohary, 2018, p. 7.

[20] Theohary, 2018, pp. 7–8.

[21] Office of the Director of National Intelligence, *Assessing Russian Activities and Intentions in Recent U.S. Elections: The Analytic Process and Cyber Incident Attribution*, ICA 2017-01D, January 2017a.

[22] Guy Taylor, "State Department Global Engagement Center Targets Russian Propaganda, 'Deep Fakes,'" *Washington Times* via Associated Press, December 12, 2018.

[23] U.S. Department of State, "About Us—Global Engagement Center," webpage, undated-a.

private industry, and others who work to study and counter state-based disinformation.[24]

These measures, however, are inputs—not outputs—and they do not capture whether these efforts are effective. Nearly every interviewee across the interagency mentioned challenges associated with measuring the effectiveness of their efforts. Many agencies use social media metrics (clicks, likes, etc.) as indicators of effectiveness, but these are not strong measures of actual influence on audience attitudes or behaviors.[25] They can offer insight into the level of an audience's exposure to content or the reach the content has achieved, but exposure and reach are not reliable measures for audience attitudes or behavior. Furthermore, social media metrics can be manipulated by sponsored trolls or by using automated agents, such as bots. In sum, the USG cannot say definitively whether these efforts are changing minds.

The interviews also highlight that USG efforts remain fractured and disjointed.[26] By law, the GEC should coordinate this effort. In practice, the GEC has liaison officers from the combatant commands, intelligence community, USAGM, and the U.S. Agency for International Development, but these officers sometimes lack the authority to speak on behalf of their parent agency and the access to senior policymakers who can.[27] Even the GEC's ability to speak authoritatively for the Department of State is questionable. As some interviewees noted, the Department of State's own organizational culture privileges the geographic bureaus as true foreign policymakers and relegates the

---

[24] Oren Dorell, "State Department Launches $40 Million Initiative to Counter Russia Election Meddling," *USA Today*, February 2018; U.S. Department of State, Office of the Spokesperson, "State-Defense Cooperation on Global Engagement Center Programs and Creation of the Information Access Fund to Counter State-Sponsored Disinformation," media note, February 26, 2018.

[25] Interview with academic, Pittsburgh, Pa., January 18, 2019.

[26] U.S. Advisory Commission on Public Diplomacy, *2018 Comprehensive Annual Report on Public Diplomacy & International Broadcasting*, Washington, D.C.: U.S. Department of State, 2018, p. 8.

[27] U.S. Department of State, undated-a.

functional bureaus—such as the GEC—to supporting and subordinate roles in the organizations.[28]

At the end of the day, who in the USG is responsible the for countering foreign influence in the information space? In short, everyone and no one. Multiple different agencies have equities in this space, but since the dissolution of the USIA, no one entity owns the problem, and even the coordination mechanisms are nascent at best.

Department of Defense Capabilities Still Nascent

Like the broader USG, the joint force is still feeling its way through what role, if any, it plays in the global counter-disinformation fight and precisely who within the joint force should have this potentially global mission.[29] The Joint Staff J-39, Deputy Director for Global Operations, is attempting to bring global fires, Information Operations (IO), and targeting efforts together to cohesively counter such campaigns.[30] The joint staff also has the Joint Information Operations Warfare Center, under the Joint Staff Director for Operations (J-3), which provides assessment tools, processes, and methodologies to help assess operations in the information environment (OIE) and oversees the development and demonstration of IO programs, including social media analysis.[31] Despite these efforts, much of the counter-disinformation fight on social media remains ad hoc and piecemeal across the joint force: Combatant commands are focused primarily on detection, consumption, and—in the case of cyber command (CYBERCOM)—production countermeasures,

---

[28] Interview with Department of State official, GEC, Washington, D.C., February 27, 2019; interview with Department of State official, Office of Public Affairs and Public Diplomacy, by phone, March 25, 2019.

[29] Jim Garamone, "Global Integration Deserves More Attention, Selva Says," U.S. Department of Defense, June 18, 2019.

[30] Interview with joint staff personnel, by phone, January 24, 2019.

[31] Chairman of the Joint Chiefs of Staff, *Charter of the Joint Information Operations Warfare Center*, Washington, D.C., CJSCI Instruction 5125.01, Washington, D.C., September 30, 2015.

and the services are focused mostly on consumption and institution-building measures (see Box 3.3).

Office of the Secretary of Defense

The fiscal 2018 National Defense Authorization Act required DoD to integrate IO across the department.[32] Mirroring the GEC's original focus on counterterrorism (rather than interstate competition), DoD gave this task to the Deputy Assistant Secretary for Defense Special Operations and Combatting Terrorism, within the Office of Secretary of Defense.[33] DoD also established a working group to manage the resources, coordinate with GEC, update the Strategy for Operations in the Information Environment and develop policy recommendations to "integrat[e] strategic information operations and cyber-enabled information operations" across the department.[34] As of September 2019, the strategy has yet to be published, and many of the more-ambitious recommendations—such as creating Under Secretary of Defense for Information or Information Command—require new funding and positions.[35] To date, none of these recommendations has been approved or decided on.

The Combatant Commands

Each of the 11 combatant commands, seven geographic combatant commands, and four functional combatant commands have the primary responsibility to counter adversary influence activities in their respective area of responsibilities.[36] In practice, the combatant commands have focused on production, detection, and consumption countermeasures.

---

32 Pub. L. 115-91.

33 Pub. L. 115-91.

34 Interview with Joint Information Operations Warfare Center personnel, by phone, August 12, 2019.

35 Interview with Office of the Secretary of Defense personnel, by phone, April 8, 2019; interview with Joint Information Operations Warfare Center personnel, by phone, August 12, 2019.

36 In practice, space command's challenges are very different from the other geographic combatant commands.

Box 3.3
Who in DoD Does What

Production [prevent actors from producing or ordering production of content]	**Distribution** [restrict actors from distributing content]	**Consumption** [build audience resilience, lower susceptibility to content]
• Attacks on producers of disinformation: CYBERCOM denial-of-service attack on IRA during 2018 elections	• N/A	• Counter messaging to foreign audience: Central Command (CENTCOM)'s Web Operations Branch (WebOps) targeting of terrorist recruitment on social media • Training service members to recognize disinformation done mostly by the services
Detection and Awareness-Raising		
• Detection: Joint Force intelligence and IO monitoring at the geographic combatant commands		
Institution-Building		

• Combatant commands: Transformation of CENTCOM WebOps to Special Operations Command's Joint Military Information Support Operations (MISO) Center
• Services: All the services are building IO capabilities.
• Army rebranded to create 1st Information Operations Command (Land)
  • U.S. Marine Corps (USMC) built out entire Marine Information Groups with a three-star Deputy Commandant for Information
  • USAF tiger team and creation of 14F field		

NOTE: Based on publicly available information.

On the production side, according to news reports, CYBERCOM conducted a denial-of-service attack against Russia's IRA—one of the organizations believed to be behind the Russian social media interference campaign in the 2016 election—in advance of the 2018 election.[37] The effort, arguably, was an example of deterrence by denial—dissuading Russian interference by increasing the costs to conduct such operations. Whether the effort accomplished this objective, however, is a matter of debate. According to Thomas Rid, a professor of strategic studies and cyber expert at Johns Hopkins University, "Such an operation would be more of a pinprick that is more annoying than deterring in the long run."[38]

While such counterproduction efforts have so far fallen mostly to CYBERCOM, many combatant commands play some role in detecting and mitigating the consumption of disinformation on social media. European Command, for instance, has incorporated real-world counter-disinformation tasks as part of military training exercises. During Trident Juncture in 2015, for instance, NATO commanders and specialists developed social media applications on the exercise's internal network to train service members on how to quickly produce high volumes of pro-NATO content through official accounts on social media to counter anti-NATO messaging.[39] Later exercises incorporated the NATO Strategic Communications Centre of Excellence [StratCom CoE] as part of the red team "to test just how much they could influence soldiers' real-world actions through social media manipulation."[40] Ultimately, this training proved particularly timely; during the subsequent Trident Juncture 2018, Russia actively targeted the exercise with an extensive disinforma-

---

[37] Ellen Nakashima, "U.S. Cyber Command Operation Disrupted Internet Access of Russian Troll Factory on the Day of 2018 Midterms," *Washington Post*, February 27, 2019.

[38] Nakashima, 2019.

[39] Gregory M. Tomlin, "#SocialMediaMatters: Lessons Learned from Exercise Trident Juncture," *Joint Force Quarterly*, No. 82, July 1, 2016.

[40] Issie Lapowsky, "NATO Group Catfished Soldiers to Prove a Point About Privacy," *Wired*, February 18, 2019.

tion campaign and NATO forces—including U.S. marines—needed to respond in real time.[41]

CENTCOM's WebOps is perhaps the most sophisticated effort by a combatant command to both detect and then counter consumption of disinformation on social media. Created in 2009, WebOps is part of CENTOM's Information Operations Division. It was established primarily to counter terrorist recruitment and propaganda activity on social media and has steadily grown in capability and personnel.[42] WebOps influences target audiences in the area of responsibility through the online information environment to counter adversary narratives and shape online environments through multiple platforms, capabilities, and technologies. It also tries to counter disinformation campaigns.

But WebOps taking on a global scope is unlikely to solve the question of how the joint force should combat disinformation campaigns on social media. First, there are the general questions of whether WebOps—which was originally designed as a counterterrorism tool for the Middle East—can scale into an organization that can confront adversaries all over the globe. Second, WebOps already faces a series of funding and personnel constraints. If WebOps mandate expands, the global scope will only exacerbate these personnel shortfalls.

The Services

The task of recruiting, organizing, and training personnel to combat disinformation on social media—i.e., the institution-building efforts as depicted in Box 3.3—falls to services. Like the combatant commands, each of the services have taken different approaches to the topic. In 2018, at the direction of Secretary Heather Wilson and Vice Chief of Staff General Stephen Wilson, USAF created a tiger team to develop concepts and solutions to address USAF deficiencies for OIE.[43]

---

[41] Donara Barojan, "#PutinAtWar: Disinformation targets Trident Juncture," *DFRLab* via StopFake.org, November 8, 2018b.

[42] Associated Press, "Report: CENTCOM Botches Effort to Fight Online Recruiting by Islamic State," *Tampa Bay Times*, January 31, 2017.

[43] Interview with USAF personnel, Washington, D.C., February 26, 2019.

In this effort, the tiger team led several working groups and conducted an innovation sprint to help USAF figure out needed capabilities and training methods to effectively engage in OIE. As of this writing, the team's effort is still ongoing, and the long-term impact of its recommendations is yet unknown.

USAF has taken some concrete steps to build a more robust IO force. In 2015, it created a new career field, designated as 14F, Information Operations Officers.[44] There are approximately 100 14Fs, mostly lieutenants and captains and mostly placed at the Air Operations Centers. [45] USAF is also building additional courses to train their IO officers on social media tradecraft,[46] and some limited reachback capability for social media analytics is conducted through the 67th Operations Support Squadron of the 24th Air Force.[47] Previously this squadron was a full wing, but it was downsized in 2010 and the IO billets were converted into cyber billets for another unit, so now only 25 personnel are dedicated to IO reachback.[48]

Despite these advances, USAF does not yet have a robust capability in this area. Given that the career field is still relatively new, most of the officers assigned to the 14F field are relatively junior. Some USAF officers are concerned that the training is focused too much on teaching how to use the tools (referred to as "buttonology") and not enough on the methodology behind the tools.[49] As for the reachback capability, the 67th still struggles in terms of personnel and resources, especially because their mandate includes other missions, such as opera-

---

[44] USAF, *Military Information Support Operations (MISO)*, Air Force Instruction 10-702, Washington, D.C., June 7, 2011.

[45] USAF, 2011.

[46] Interview with USAF personnel, Washington, D.C., February 26, 2019.

[47] Interview with 67th Air Operations Support Squadron personnel, by phone, January 18, 2019.

[48] This transition was part of a broader push with the standing up CYBERCOM to build cyber capabilities inside USAF. U.S. Air Forces Cyber, *Gunslingers: A Brief History of the 67th Cyberspace Wing*, undated; interview with 67th Air Operations Support Squadron personnel, by phone, January 18, 2019.

[49] Interview with Air Force A2 service member, Washington, D.C., February 27, 2019.

tional security. The unit has mostly focused on sifting through publicly available information to see whether USAF aircraft have been reported leaving or arriving at airfields, commonly referred to as *tail spotting*.[50] The unit lacks both the resources and the training to look for disinformation campaigns, nor does it have a clear mechanism for passing the information along to another element that can investigate further.[51]

By contrast, the Army has the greatest capability to engage on social media platforms. The lead organization for these efforts is the U.S. Army Cyber Command—particularly, its 1st Information Operations Command (Land) which provide IO and cyberspace operations support through deployable teams, reachback support and specialized training.[52] The Army Information Operations Center within 1st IO provides reachback capability, intelligence analysis, and technical assistance to deployed personnel and military units.[53] The center does provide internal training that focuses on social media analysis and collection, but this is considered a specialized skill and requires more than six months to effectively train an analyst to understand the tools, the techniques, and the signs to look for on social media.[54]

The Army's ability to influence foreign audiences comes primarily from two active duty 4th and 8th Military Information Support Groups both of which increasingly use social media to conduct influence operations.[55] Much of the Army's MISO capability, however, rests in the reserve component. The U.S. Army Civil Affairs and Psychological Operations Command (Airborne) provides 83 percent of DoD's

---

50 Interview with 67th Air Operations Support Squadron service member, by phone, January 18, 2019.

51 Interview with 67th Air Operations Support Squadron service member, by phone, January 18, 2019.

52 U.S. Army Cyber Command, "History," webpage, undated-b.

53 U.S. Army Cyber Command, "1st Information Operations Command (Land)," webpage, undated-a.

54 Interview with 1st Information Operations Command personnel, by phone, July 9, 2019.

55 Interview with Army psychological operations personnel, Fort Bragg, July 9, 2019.

psychological operations forces.[56] Like their active-duty counterparts, these forces also focus on psychological operations social media, although integrating this capability into conventional military training exercises, such as national training center rotations, historically has proven challenging.[57]

Though not as large numerically as the Army, the USMC has developed innovative solutions to building information capabilities into its force. At the headquarters level, the USMC has a three-star Deputy Commandant for Information responsible for developing plans, policies, and strategies for OIE.[58] On the operational level, the USMC created the Marine Information Group. This subcomponent to the Marine Expeditionary Force, led by a colonel reporting directly to the force commander, features capabilities for MISO, public affairs and combat camera (now merged to become communications strategy and operations), intelligence, and other specialties to help the Marine Expeditionary Force contest activities in the information environment.[59] Finally, the USMC has the Marine Corps Information Operations Center, which provides operational support to USMC components by supplying IO subject-matter expertise.[60]

Finally, the Navy is arguably the least engaged in the counter-disinformation fight on social media. Most of its capabilities to engage and monitor social media are conducted through intelligence and communications offices.[61] Specifically, the Navy National Maritime Intelligence Center has a limited ability to monitor and analyze social

---

[56] U.S. Army Reserve, "U.S. Army Civil Affairs & Psychological Operations Command (Airborne)," webpage, undated.

[57] Interview with Army CYBERCOM and IO personnel, by phone, July 9, 2019.

[58] Mark Pomerleau, "Why the Marine Corps Needed a New Deputy Commandant," *C4ISRNET*, December 5, 2017.

[59] Deputy Commandant for Information, "Brief: MAGTF Operations in the Information Environment (OIE)," April 9, 2019.

[60] USMC, "MCINCR—Marine Corps Base Quantico," webpage, undated.

[61] Mark Pomerleau, "How the Navy Is Changing Its Thinking on Information Warfare," C4ISRNET, April 21, 2019.

media platforms.[62] Beyond the National Maritime Intelligence Center, however, much of the Navy has relatively little interaction with social media analysis—and because the Navy does not have an occupational code for these skills, it has relatively few trainers and analysts to educate the Navy on these process and techniques.[63]

Recently, the Navy has signaled some interest in developing a more robust IO capability, changing the name of the information dominance career field to a broader career field titled information warfare.[64] This will unite such fields as cryptology, oceanography, space, information, and intelligence to better train and operate in the information space.[65] Similarly, the Navy also renamed its Space and Naval Warfare Systems Command to the Naval Information Warfare Systems Command.[66] Still, the service lags behinds its counterparts.

In the end, how each service builds IO capabilities can have second-order effects. Most directly, it shapes how forces are presented to the combatant commands for employment. Some services—such as the Army and USMC—can offer entire units; others might need to rely on one-off augmentees. Less directly, these actions can also shape how the services administer training for this capability.

All the services struggled to train service members on how to combat disinformation campaigns, particularly inside the United States. Part of the challenge is legal: The U.S. military is prohibited by law from conducting IO inside the United States.[67] Another obstacle is

---

[62] Interview with National Maritime Intelligence Center personnel, Maryland, March 14, 2019.

[63] Interview with National Maritime Intelligence Center personnel, Maryland, March 14, 2019.

[64] Henry Stephenson, "Navy Information Warfare: A Decade of Indulging a False Analogy," *Proceedings*, U.S. Naval Institute, Vol. 145, No. 1/1,391, January 2019.

[65] Robert Ackerman, "The Navy Turns a Sharper Eye Toward Information Warfare," *Cyber Edge*, February 1, 2019.

[66] Naval Information Warfare Systems Command Public Affairs, "SPAWAR Changes Name to Naval Information Warfare Systems Command—Aligns Identity with Mission," June 3, 2019.

[67] Pub. L. 110-417.

institutional. Training to combat disinformation campaigns on social media—and MISO more broadly—are usually viewed as second-tier objectives when integrated into large-scale conventional training objectives. Although psychological operations are part of an Army brigade's rotations to the national training centers, for example, they are not the focus of these rotations, and disinformation on social media is often relegated to a notional event.[68]

More recently, the IO community has developed some tools for training in a simulated fashion. For example, the Information Operations Network (ION), a stand-alone replication of the internet, allows service members to train on social media analysis and collection without having to worry about crossing legal boundaries.[69] ION also allows exercise controllers to turn up or down the level of *noise*—unrelated information placed onto the system to complicate the analysis. Since its start in 2014, ION supported more than 150 exercises by the Army, USAF, and USMC, and it was used by all the large-scale training centers.[70] Nonetheless, even ION cannot fully replicate the vastness and complexity of the actual information environment.

Conclusion: A Still Nascent Capability

Ultimately, the joint force has yet to develop a fully coherent approach for conducting a counter-disinformation fight. Pieces of a potential solution are in place. CENTCOM's WebOps, for example, provides a starting point for how to counter certain types of disinformation. The USMC probably has the best approach for fielding trained forces ininformation warfare, having designated a three-star proponent at the headquarters level, developed operational officers, and built robust reachback capability. And Trident Juncture shows how disinformation on social media can be incorporated into conventional military training.

---

[68] Interview with Army CYBERCOM and IO personnel, Washington, D.C., July 9, 2019.

[69] U.S. Army Training and Doctrine Command, "G-2 Operational Environment Center TRADOC G-2: Information Operations Network," webpage, undated.

[70] Interview with ION personnel, Fort Belvoir, July 8, 2019.

Despite these high points, much of the joint force's effort still seems piecemeal and ad hoc. CENTCOM's WebOps, at least for the moment, remains focused on a problem set in one part of the world and not on the full range of U.S. adversaries spanning the globe. The USMC might have structures in place, but other services are still wrestling with how they will tackle this problem set for themselves and how they will build forces for the combatant commands to employ. And although Trident Juncture provides one example for how to train for disinformation on social media, it is still just one set of exercises.

Allies and Partners Have Diverse Contexts, Varied Approaches

The USG is not alone in wrestling with how to deal with disinformation on social media. Countries around the world are dealing with similar problems. And as is the case within the United States, other approaches have been varied and often ad hoc in nature. Perhaps because the problem itself is still comparatively new, countries are still experimenting with a wide variety of solutions—targeting all five of the categories of countermeasures noted in Box 3.4 with varying degrees of intrusiveness and each with mixed effects.

Message Discipline

Perhaps the least intrusive way for a government to combat disinformation is enforcing message discipline and tamping down its own misinformation. During the height of the fighting in Eastern Ukraine, for example, Russian disinformation efforts would capitalize on the Ukrainian government's conflicting accounts of the fighting and casualties to depict the Ukrainian government as incompetent and the situation as more dire than it was in reality. Thus, Ukraine's first step in combating Russian disinformation was for the Ukrainian government to get its own version of events straight.[71]

---

[71] Interview with think tank analyst, Kyiv, Ukraine, March 5, 2019.

In Ukraine's case, the government and military implemented a "One Voice Policy" to ensure that the government was putting out only one narrative.[72] The heads of each agency would get together daily, decide on the facts they would all disseminate, and hold a daily official press conference. [73] Additionally, the Ukrainian military designated a handful of selected spokespersons to go on television and explain what was occurring at the front lines.[74]

Obviously, other countries face situations that are less dire than what Ukraine faced in 2014 and 2015, but the underlying thesis that message discipline—and, more broadly, minimizing the opportunities for foreign disinformation efforts to play off half-truths—remains broadly applicable.

Institution Building and Media Education:
Beyond enforcing message discipline across governments, some states have set up entire institutions and built out specific programs to counter this disinformation. Like in the United States, the response to disinformation on social media rarely falls squarely on a single ministry or government entity. As a result, Sweden, Finland, Denmark, the United Kingdom, the Netherlands, and Latvia have organized networked institutions for this purpose—bringing together multiple ministries (often including the foreign ministry, defense ministry, security services and civil response organizations). The Czech Republic has also set up centers devoted to the threat.[75] Ukraine established a separate Ministry of Information Policy of Ukraine on December 14, 2014, to "develop strategies for information policy of Ukraine and the concept of information security"; to "coordinate government agencies in mat-

---

[72] Interview with think tank analyst, Kyiv, Ukraine, March 5, 2019; interviews with midgrade military officers, Kyiv, Ukraine, March 7, 2019.

[73] Interview with think tank analyst, Kyiv, Ukraine, March 5, 2019; interviews with midgrade military officers, Kyiv, Ukraine, March 7, 2019.

[74] Interview with think tank analyst, Kyiv, Ukraine, March 5, 2019.

[75] Naja Bentzen, "Foreign Influence Operations in the EU," European Parliamentary Research Service, July 2018, p. 7; Vilmer et al., 2018, p. 117.

Box 3.4
Actions Taken by U.S. Allies and Partners

Production [prevent actors from producing or ordering production of content]	**Distribution** [restrict actors from distributing content]	**Consumption** [build audience resilience, lower susceptibility to content]
• Pacts: German party agreement not to use disinformation in advance of 2017 election (making such attacks less profitable to conduct) • Threats: German threats to Russia in advance of 2017 elections	• Code of conduct: Japan's voluntary pact commits social media companies to minimize spread of disinformation • Regulation: Singapore's Protection from Online Falsehoods and Manipulation Law bans bots and allows for content removal • Regulation: South Korea's National Security Act (1948) criminalizes the possession, access, and dissemination of North Korean propaganda	• Message discipline: Ukrainian attempt to enforce message discipline during early phases of 2014–2015 conflict and proactively reduce opportunities for disinformation • Media literacy program: National-level effort to train populations to spot disinformation • Debunking: EU StratCom East's EU vs. Disinfo • Regulation: Singapore's Protection from Online Falsehoods and Manipulation Law demanding that social media companies post info to debunk falsehoods • Bans: Ukraine ban on cellphones for soldiers serving in the Donbass; later ban on VK
Detection and Awareness-Raising • Detection: Lithuania's development of an artificial intelligence–enabled detection system; Swedish Civil Contingencies Agency's cooperation with social media; EU's Rapid Alert System; NATO StratCom COE		
Institution-Building • Institution-building: New intergovernmental organizations (EU StratCom East; NATO StratCom CoE); national-level efforts across Europe and Asia		

NOTE: This list is illustrative rather than comprehensive.

ters of communication and information dissemination"; and, finally, to "counteract informational aggression by Russia."[76]

The common thread running through many of these organizations is media education. Finland launched an extensive counter-disinformation media literacy campaign in 2014, well before the 2016 Russian election interference.[77] Canada, Australia, and Sweden incorporated media literacy training into their youth education systems.[78] Similarly, Ukraine's Ministry of Information Policy has had a hand in identifying Russian disinformation campaigns, providing civic education about how to spot Russian disinformation, and building its own information campaigns.[79] The ministry also tried to tell Ukraine's story to Western audiences, albeit with mixed success.[80] In all these cases, assessing the efficacy of these efforts in increasing resilience for foreign disinformation proves challenging.

In Singapore, the government began an information literacy campaign called S.U.R.E. (Source, Understand, Research, and Evaluate) in 2013 with the goal of educating primary and secondary school students about how to evaluate content trustworthiness, including a section on how to recognize fake news.[81] And yet, like other public education efforts, it is difficult to quantify the success of such efforts.

---

[76] Minister of Information Policy of Ukraine, "About the Ministry," webpage, undated.

[77] Eliza Mackintosh, "Finland Is Winning the War on Fake News. What It's Learned May Be Crucial to Western Democracy," CNN, May 2019. There is an open question about the extent to which the Finnish example—with a small, relatively homogenous population situated next to a larger, long-standing adversary—is generalizable to other countries.

[78] Dana Priest and Michael Birnbaum, "Europe Has Been Working to Expose Russian Meddling for Years," *Washington Post*, June 25, 2017.

[79] Interview with politician, Kyiv, Ukraine, March 5, 2019.

[80] Interview with senior government official, Kyiv, Ukraine, March 6, 2019.

[81] National Library Board (Singapore), "S.U.R.E. Campaign," webpage, undated-b; National Library Board (Singapore), "Fact-Checking Using Multiple Sources," webpage, undated-a.

Detection

Countries have also tried to ramp up efforts to detect state-sponsored disinformation. In some cases, this work has fallen to intelligence agencies. During the 2017 German election, the German domestic intelligence agency BfV shared information with political parties on potential threats.[82] Other countries have pursued innovative public-private partnerships to aid their detection efforts. For example, the Swedish Civil Contingencies Agency cooperated with social media companies in efforts to better detect and understand Russian IO on their platforms.[83] Still others have turned to new technology for a solution. The Lithuanian defense ministry in 2018 claimed to have invented an artificial intelligence program that flags disinformation within two minutes of its publication and sends those reports to human specialists for further analysis.[84]

Intergovernmental organizations, particularly in Europe, have also taken on some of the detection function. The EU Intelligence and Situation Center has tried to facilitate the exchange of information across member states to better detect Russian disinformation attempts.[85] The EU in 2019 set up a Rapid Alert System, a common information-sharing platform created to "facilitate the sharing of data and assessments of disinformation campaigns and to provide alerts on disinformation threats in real time."[86] Similarly, in 2014, NATO created the StratCom CoE in Riga, Latvia, to improve NATO members' understanding of hostile

---

[82] Erik Brattberg and Tim Maurer, *Russian Election Interference: Europe's Counter to Fake News and Cyber Attacks*, Washington D.C.: Carnegie Endowment for International Peace, May 23, 2018, footnote 105.

[83] Dominik Swiecicki and Irene Christensson, "MSB Activities on Countering Information Influence Campaigns," Counter Influence Branch, Global Monitoring and Analysis Section, Swedish Civil Contingencies Agency, presentation at RAND Corporation in Santa Monica, Calif., November 29, 2018.

[84] Iryna Somer, "Lithuanians Create Artificial Intelligence with Ability to Identify Fake News in 2 Minutes," *Kyiv Post*, September 21, 2018.

[85] Vilmer et al., 2018, p. 134.

[86] European Commission Press Release, "A Europe That Protects: The EU Steps Up Action Against Disinformation," press release, December 5, 2018.

information efforts, including those distributed via social media.[87] As of 2018, the CoE had a team of roughly a dozen people dedicated to detecting disinformation operations.[88] Despite its small size, the CoE has notched some noteworthy successes. For example, after suspected Russian actors planted a fake report about German soldiers stationed in Lithuania raping a local girl, Lithuanian communications specialists flagged the report for other NATO members.[89]

Debunking

After disinformation is detected, the question becomes how to minimize its effects. The most common answer, perhaps, has been to debunk the false or misleading claims. Much of the effort in this domain is routine: States regularly respond to what they consider false claims via press statements or in senior leader interviews with the press. For instance, the Singaporean government set up a website called Factually that aims to clarify widespread or common misperceptions of policies or other matters of public concern.[90] Especially for those states that are routinely targeted by foreign disinformation campaigns, such as Taiwan, this can be a time-consuming endeavor. One Taiwan official mentioned being forced during the president's trip to respond to a grand total of seven false stories in a single day, consuming an enormous amount of time and resources on the part of senior government officials.[91]

Some intergovernmental initiatives are worth noting. In 2015, the EU created the East StratCom Task Force to debunk disinformation (primarily of Russian origin); promote a positive message about the EU; and to some extent, support professional media in Eastern Euro-

[87] NATO Strategic Communications Centre of Excellence, "About Strategic Communications," webpage, undated.

[88] Vilmer et al., 2018, p. 135.

[89] Deutsche Welle, "Russia's Information Warfare Targets German Soldiers in Lithuania," Atlantic Council webpage, February 24, 2017.

[90] Government of Singapore, Factually website, undated. Kelly Ng, "The Big Read: In an Era of Fake News, the Truth May Not Always Be Out There," *Today*, June 2, 2017.

[91] Interview with senior government official, Taipei, Taiwan, January 2019.

pean countries.[92] This task force often partners with local civil society groups that track Russian disinformation and publicizes debunked material on the EU vs. Disinfo website and on social media as EU Mythbusters.[93]

Even seemingly benign actions, such as debunking disinformation, can still prove controversial. There is an open question about whether debunking does more harm than good, especially when it attracts more attention to an otherwise obscure story. Moreover, in the case of intergovernmental efforts, calling out disinformation risks trampling on domestic political sensitivities, especially if it is produced by local actors inside the European Union members rather than by foreign actor like Russia.

Pacts and Codes of Conduct

Aside from debunking, some states and intergovernmental organizations have pushed for pacts or codes of conduct. Most of these codes of conduct have centered on stemming the spread of disinformation. For example, the European Commission pushed an EU-wide "Code of Practice on Disinformation," which commits signatory social media platforms to implement a variety of countermeasures, such as closing fake accounts and identifying bot-spread content.[94] Similarly, in January 2019, Japan planned to develop "a set of measures aimed at preventing the spread of false online information . . . particularly during elections and disasters."[95] The plan had yet to be formalized when this report was written, but press accounts suggested that it could "include requesting that major U.S. technology companies and other information providers voluntarily formulate a code of conduct," and that it might call on Japan-based tech companies, such as LINE Corporation (owned by a South Korean company) and Yahoo Japan Corporation

---

[92] Interview with European Union officials, by telephone, November 29, 2018.

[93] Interview with European Union officials, by telephone, November 29, 2018.

[94] European Commission, 2018.

[95] Kyodo, "Japan Plans to Take Steps Against 'Fake News' by June," *Japan Times*, January 14, 2019.

(owned by a Japanese company), to improve measures for combating the spread of misinformation on their platforms.[96]

In some countries, these pacts have gone one step further—with potential beneficiaries of disinformation promising not to use it to their advantage. For example, during the 2017 election, the German political parties reached a "gentlemen's agreement" not to use bots on social media or exploit any hypothetical leaked information.[97]

Whether or not these voluntary pacts or codes of conduct could be successful outside the unique contexts of Germany and Japan, however, remains open to discussion. In 2017, the major German political parties still largely held to "the pro-EU and internationalist consensus" and agreed about the key issues Russia cared about—such as the EU, NATO, and the Ukraine crisis—making it an unattractive target for Russian disinformation. [98] Even if more disinformation existed, the parties might not have seen it as particularly politically useful, because Germans "unlike Americans . . . [tend to be] wary of information disseminated on Facebook and Twitter."[99] Similarly, although the strength of any voluntary pact in Japan remains hypothetical at this point, Japan is a relatively cohesive society with a strong reliance on print—rather than social—media and a relatively dominant state-owned media outlet.[100]

Laws and Regulations

Instead of voluntary pacts, some states have tried to find legislative solutions for disinformation on social media. Singapore stands as the foremost example in this realm. In 2019, Singapore passed the Protection from Online Falsehoods and Manipulation Law, which gives the govern-

---

[96] Kyodo, 2019.

[97] Brattberg and Maurer, 2018, p. 18.

[98] Tyson Barker, "Germany Strengthens Its Cyber Defense: How It's Meeting the Russian Threat," *Foreign Affairs*, May 26, 2017.

[99] Michael Schwirtz, "German Election Mystery: Why No Russian Meddling?" *New York Times*, September 21, 2017; Barker, 2017.

[100] Interview with Japanese government official, January 16, 2019; interview with Japanese political and defense analyst, January 2019.

ment (but not individual users) the ability to request online platforms—whether they are traditional media or social media—to post corrections of statements that the government deems demonstrably false and against the public interest or to remove those posts if corrections are not issued.[101] The law also bans the use of fake online accounts and bots. If the platform refuses to issue corrections or remove the false posts, the government can block the website or take it to court.[102] Failure to comply can bring fines and imprisonment, but only if the transgressor knowingly shares false content.[103] The law even applies to closed platforms, including online chat groups, such as LINE, and applications with end-to-end encryption, such as WhatsApp, although it is unclear how Singapore will be able to enforce such a rule.[104]

Singapore's law has proven controversial. Proponents claim that the law is meant to cover statements of fact, not academic discourse, opinions, criticism, satire, or parody.[105] Singapore-based academics, journalists, and tech companies counter that the law was, at best, unnecessary and, at worst, a bid to curtail freedom of speech and expression, giving the government an enormous power to decide what information is true or false and thus what gets taken down, blocked, or corrected.[106]

---

101 Parliament (Singapore), Protection from Online Falsehoods and Manipulation, Bill Number 10/2019, April 1, 2019; interview with defense and political analyst, Singapore, May 2019. For a good summary of the key points of the bill that eventually became law, see Lim Min Zhang, "Fighting Fake News Here with Legislation," *Straits Times*, May 13, 2019.

102 Singapore government official, email with authors, June 21, 2019.

103 Interview with political and defense analyst, Singapore, May 2019; Singapore government official email, 2019.

104 Of note, it is unclear whether Singapore can enforce this part of the regulation. Interview with government official, Singapore, May 2019; interview with political and defense analyst, Singapore, May 2019.

105 Cara Wan, "No Need to Be Overly Worried About Fake News Laws, Says Ong Ye Kung," *Straits Times*, April 29, 2019.

106 Kirsten Han, "Why Singapore's Moves to Curb 'Fake News' May Backfire," *Washington Post*, March 5, 2018; Tessa Wong, "Singapore Fake News Law Polices Chats and Online Platforms, BBC, May 9, 2019; Hillary Leung, "Singapore Is the Latest Country to Propose Tough Legislation Against Fake News," *Time*, April 2, 2019; Michelle Toh, "Google Says Singapore

Outside Singapore, other legislative initiatives are underway. South Korea, through its National Security Act (also known as the National Security Law) adopted in 1948, bans access to North Korean propaganda media and content and criminalizes the possession, access, and dissemination of such content.[107] North Korean propaganda websites, such as Uriminzokkiri, are therefore blocked within South Korea. However, the sweeping provisions of this law have been widely abused throughout history to prosecute regular domestic political dissidents and suppress freedom of expression. NGOs, such as Human Rights Watch, have consistently called for the repeal of this law,[108] and although it has been applied more judiciously in recent years, no institutional restraints prevent a return to a broader application of the law or its extension to social media accounts and content.

In Europe, Germany likely has the most-aggressive laws regulating hate speech. Its *Netzwerkdurchsetzungsgesetz*, sometimes called the NetzDG law or internet transparency law, requires social media companies to "remove posts that contain hate speech or incite violence within 24 hours or face fines as high as €50 million."[109] In July 2019, Germany fined Facebook €2 million (or $2.3 million) for violating this law.[110]

The EU's General Data Protection Regulation (GDPR), which was passed by the European Parliament in 2016 and came into effect in 2018, was not intended as counter-disinformation regulation.[111] Still, by mandating data privacy protections and imposing hefty fines for failure to comply, GDPR makes personal data more difficult to

Risks Hurting Innovation with Fake News Law," CNN, May 9, 2019; "Singapore Fake News Law a 'Disaster' for Freedom of Speech, Says Rights Group," *The Guardian*, May 9, 2019.

107 Government of South Korea, National Security Act, Korea Law Translation Center, 2011.

108 Human Rights Watch, "South Korea: Cold War Relic Law Criminalizes Criticism," May 28, 2015.

109 "Germany Fines Facebook for Underreporting Hate Speech Complaints," *DW*, July 2, 2019.

110 "Germany Fines Facebook . . .," 2019.

111 GDPR, "GDPR FAQs," webpage, undated-a.

acquire—and, by extension, makes it more difficult to microtarget individuals using disinformation that is based on their personal characteristics.[112] GDPR remains a relatively new legislative initiative, however, and time will tell whether it actually proves effective as a disinformation countermeasure.

Bans

Perhaps the most draconian defense countermeasure has been to ban social media use altogether. Obviously, few countries have adopted such measures, but given the existential nature of the threat, Ukraine experimented with banning to a limited degree. The military banned soldiers' use of cellphones on the front lines for a mixture of operational security reasons and counter-disinformation reasons.[113] According to many Ukrainian officers, such bans proved impractical to enforce, and bored soldiers still found ways to smuggle cellphones to the front lines.[114]

In an even more draconian measure, on May 15, 2017, Ukraine instituted a ban on its most popular social networking platform at the time, the Russian-owned VK.[115] Ukrainians could still access VK through a virtual private network (VPN), but the ban prevented direct Ukranian access of VK.[116] According to analysis done for NATO by the data analytics firm Singularex, Ukrainian posts dropped in half as the ban was going into effect (see Figure 3.1), presumably because many Ukrainians chose to comply with the law or because the need to use a VPN was a sufficiently large obstacle.[117] A second smaller drop in VK usage occurred several months later, from February through

---

[112] GDPR, "GDPR Key Changes," webpage, undated-b.

[113] Interview with midgrade military officers, Kyiv, Ukraine, March 7, 2019.

[114] Interview with midgrade military officers, Kyiv, Ukraine, March 7, 2019.

[115] Anton Dek, Kateryna Kononova, and Tetiana Marchenko, "The Effects of Banning the Social Network VK in Ukraine," in *Responding to Cognitive Security Challenges*, Riga, Latvia: NATO Strategic Communications Centre of Excellence, January 2019, p. 39; interview with a think tank analyst, Kyiv, Ukraine, March 5, 2019.

[116] Interview with data analytics firm, Kyiv, Ukraine, March 9, 2019.

[117] Interview with data analytics firm, Kyiv, Ukraine, March 9, 2019.

April 2018, possibly because users lost interest in VK as their friends migrated to other platforms.[118] Unsurprisingly, the ban on VK benefited other platforms; the popularity of Facebook and YouTube increased dramatically.[119]

Evidence is mixed about whether the ban proved an effective way for fighting Russian disinformation. The ban reduced Russia's access to Ukrainians' personal information, and it likely complicated Russian disinformation efforts if only because the Russians had to operate on U.S. social media platforms instead of Russian ones. However, Singularex found that Ukrainian users who chose to remain on VK after the ban tended to be younger and more ideological.[120] The number of ideological posts increased by 1.22 times after the ban went into effect, most notably in pro-Russian propaganda.[121] In other words, by pushing most of the apolitical Ukrainian user base off of VK, Ukraine might have made VK a smaller but more virulent platform for Russian disinformation than it was previously.

Threats

Finally, states have deterred disinformation campaigns by threatening and sometimes enacting various forms of retaliation. During the 2017 German elections, President Frank-Walter Steinmeier stated that, "Were Moscow to interfere in the election of the Bundestag, then the share of commonalities will necessarily decrease further. That would be damaging for both sides."[122] Whether the threat worked is open to debate, but—contrary to many experts' expectations—there was no reported Russian meddling in the elections.[123] Other states have engaged in defensive cyberoperations, imposed sanctions, penalized RT, and banned Russian-affiliated media, all in an effort to increase

---

[118] Interview with data analytics firm, Kyiv, Ukraine, March 9, 2019.

[119] Dek et al., 2019, p. 41.

[120] Dek et al., 2019, pp. 45, 57.

[121] Dek et al., 2019, pp. 48, 50.

[122] Brattberg and Maurer, 2018, footnote 99.

[123] Schwirtz, 2017.

Figure 3.1
Social Media Platform Use of VK over Time

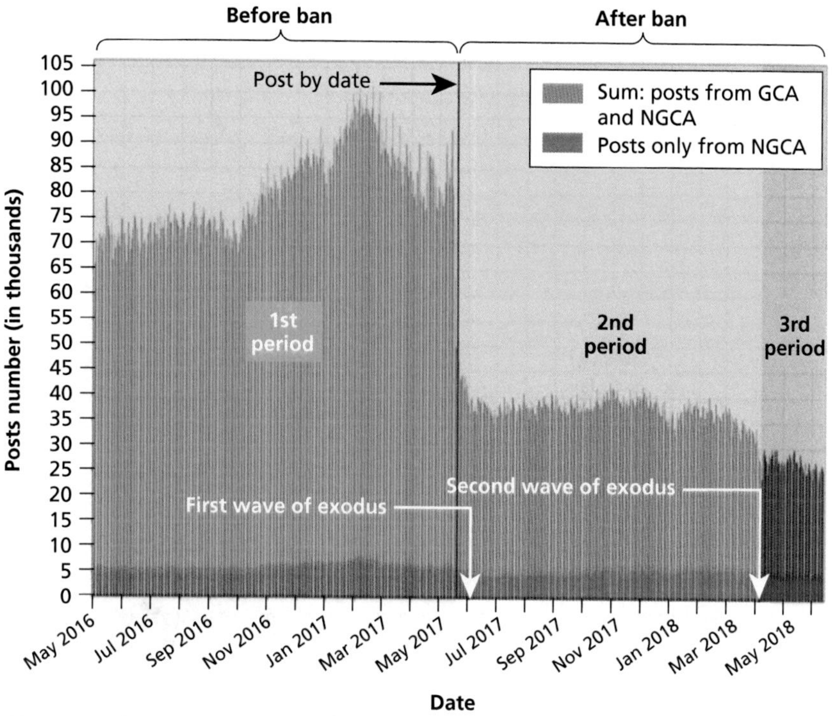

SOURCE: Dek, Kononova, and Marchenko, 2019, p. 43.
NOTE: GCA = government-controlled area; NGCA = non–government-controlled area.

the costs to Russia for these disinformation campaigns.[124] Given the power disparities between Russia and many of its smaller European neighbors in the East and between China and Taiwan, there are limits to what threats and offensive actions can accomplish.

---

124 Latvian Public Broadcasting, "Latvia Shuts Down Sputnik Propaganda Website," March 29, 2018.

Industry Limited and Motivated by Bottom Lines

For two days in April 2018, 100 lawmakers grilled Facebook founder and chief executive Mark Zuckerberg about his company's role in selling user information to the data analytics firm Cambridge Analytica and for tacitly abetting Russia's disinformation campaign in the 2016 presidential election.[125] For Facebook, the highly publicized testimony was more than a public spectacle; it was a make-or-break event that would affect its share price; regulatory environment; arguably, its very future. By all accounts, Zuckerberg fared well in his public testimony and Facebook's share price even rose over the two days, but the event vividly captured the central role that social media companies play in the disinformation fight.[126]

Perhaps more than is the case for other strategic problems, combating foreign disinformation efforts falls on the private sector. Social media companies are independent businesses, and combating disinformation has become a business imperative for many of them. As Facebook acknowledged in its 2016 Form 10-K (an annual report on the company's financial performance required by the U.S. Securities and Exchange Commission), if users do not "perceive [the platform's] products to be useful, reliable, and trustworthy," Facebook, along with other tech giants, might struggle to attract and retain users.[127] Moreover, several countries across Europe, the Americas, and Asia are considering some sort of regulation of these platforms to combat misinformation, disinformation, and other harmful content.[128] By countering disinformation, the companies could then argue that the industry could police itself.

---

125 Dustin Volz and David Ingram, "Facebook's Zuckerberg Unscathed by Congressional Grilling, Stock Rises," Reuters, April 11, 2018.

126 Volz and Ingram, 2018.

127 U.S. Securities and Exchange Commission Form 10-K, Facebook, Inc., 2016.

128 Reality Check Team, "Social Media: How Can Governments Regulate It?" BBC News, April 8, 2019; Ralph Jennings, "In the Name of 'Fake News,' Asian Governments Tighten Control on Social Media," *Los Angeles Times*, February 8, 2019.

The social media giants have taken a threefold approach to combating disinformation. First, some companies have taken steps to verify the identities of their users and remove accounts that misrepresent identities. Twitter periodically conducted large-scale purges of accounts for violating its terms of service (including spreading disinformation), deleting nearly 6 percent of its active accounts in July 2018 and then another 5,000 Iranian, Russian, and Venezuelan accounts in January 2019.[129] As noted in Chapter One, disinformation includes not only false or misleading content but also false or misleading identities, so making sure that users are who they say are is an important countermeasure in stopping the production of disinformation.

Second, technology companies have actively tried to identify and then crack down on disinformation on their platforms (or distribution and detection countermeasures as noted in Box 3.5), particularly around high-profile events, such as elections. Facebook stood up its war room in September 2018 to counter disinformation campaigns targeting the U.S. and Brazilian elections, and employed some 15,000 contractors in February 2019 to help identify questionable content and the company.[130] Microsoft started its Defending Democracy Program in March 2018 to provide an added level of cyberprotection, mostly to email accounts (rather to social media per se) for political campaigns.[131] Alphabet is working with government and law-enforcement entities to better insulate elections from the spread of disinformation over Google's platforms—specifically, Google Search, Google News, and YouTube.[132]

Second, technology companies have promoted digital literacy to boost the general public's awareness about disinformation (or con-

[129] Julia Jacobs, "In Twitter Purge, Top Accounts Lose Millions or Followers," *New York Times*, July 12, 2018; Cristiano Lima, "Facebook, Twitter Take Down Disinformation Campaigns Linked to Iran, Russia, Venezuela," *Politico*, January 31, 2019.

[130] Joshua Brustein, "Facebook Grappling With Employee Anger over Moderator Conditions," Bloomberg, February 25, 2019; Samidh Chakrabarti, "Fight Election Interference in Real Time," Facebook Newsroom, October 18, 2018.

[131] Interview with tech company representative, Seattle, Wash., February 22, 2019.

[132] Interview with tech company representative, San Francisco, Calif., April 10, 2019.

sumption and awareness-raising countermeasures from Box 3.5). For example, by 2020, Facebook intends to launch an Asia-Pacific–focused initiative called We Think Digital to raise the level of critical thinking on the company's platforms and on the internet as a whole.[133] Similarly, Twitter partnered with civil society actors, schools, the United Nations Educational, Scientific and Cultural Organization, and other organizations to promote media literacy and campaigns designed to improve disinformation awareness and detection. [134] Microsoft and Google joined with Newsguard, an organization using "journalism to fight false news, misinformation, and disinformation" by providing ratings and descriptions of news articles and websites as users browse the internet.[135]

Despite these broad generalities, the tech industry remains divided in its approach to disinformation. As mentioned earlier, foreign social media platforms, such as VK, and smaller platforms, such as 4chan or Reddit, have not been as invested in countering disinformation as their larger counterparts, possibly stemming from lack of will or resources.

Even among those companies that are actively engaged in combating disinformation, there is no consensus on best practices for identifying disinformation and what to do about it once it is detected. Twitter, for example, primarily uses automated detection mechanisms to help identify disinformation campaigns; Facebook employs not only automated systems but also extensive use of third-party fact-checkers.[136] Companies also disagree about what do with disinformation once it is found. As

---

[133] Clair Deevy, "Introducing We Think Digital: New Digital Literacy Resources to Reach 1 Million People in Asia Pacific by 2020," Facebook Newsroom, March 4, 2019.

[134] Ronan Costello, "Celebrating #EUMediaLiteracyWeek," Twitter Blog, March 20, 2019; Karen White, "Improving Health During Global Media and Information Literacy Week," Twitter Blog, October 24, 2018.

[135] Newsguard, "The Internet Trust Tool," webpage, undated.

[136] Interview with tech industry representative, San Francisco, Calif., March 21, 2019; interview with tech industry representative, San Francisco, Calif., February 6, 2019. Facebook's third-party fact-checkers can give content one of nine different ratings: false, mixture, false headline, true, not eligible, satire, opinion, prank generator, and not rated. "Third-Party Fact-Checking on Facebook," Facebook, undated.

Box 3.5
What Technology Companies Are Doing

Production [prevent actors from producing or ordering production of content]	**Distribution** [restrict actors from distributing content]	**Consumption** [build audience resilience, lower susceptibility to content]
• Verifying the identity of users and removing content and accounts (e.g., Twitter purging fake Russian, Iranian, and Venezuelan accounts)	• Standing up war rooms to detect disinformation • Downgrading content	• Digital literacy campaigns (e.g., Facebook's We Think Digital; Twitter partnering with United Nations)
Detection and Awareness-Raising		
• Detection: Use of external fact-checking groups and automated mechanisms to detect inauthentic content		
Institution-Building		
• N/A		

NOTE: This list is illustrative rather than comprehensive.

already mentioned, Twitter and Facebook have purged fake accounts. Search engines—such as Google and Bing—have tried to promote high-quality, genuine content and downgrade low-quality or misleading pages.[137] Facebook publishes pieces debunking disinformation.[138]

Approaches to disinformation can differ even within the same corporation. Alphabet polices its Google search engine differently from how it polices YouTube and Google News.[139] Facebook treats its namesake platform differently from Instagram and WhatsApp.[140] Some of these differences are inherent to platform design: A messaging platform, such as WhatsApp, faces different challenges than social networking platforms, such as Facebook. Some differences stem from history and mission. Jigsaw (formerly Google Ideas) was founded with a public-interest mission of making the internet better rather than for profit.[141]

The tech industry does face several limitations in countering disinformation. These companies are, after all, for-profit international corporations with responsibilities to their shareholders, and their interests do not always neatly align with those of the USG. From the companies' standpoint, ignoring the disinformation problem carries business risk, but taking an overly heavy-handed approach risks alienating some of their user base and could hurt profits.

Moreover, because many of these companies are global, they are wary of developing too close a relationship with any one government. What counts as disinformation can be a subjective judgment call, and counter-disinformation campaigns can easily be abused by authoritarian regimes to crack down on political opponents. As demonstrated in Figure 3.2, Twitter receives exponentially more requests to remove content from authoritarian or semiauthoritarian regimes (Russia and Turkey) than the next six countries (all liberal democracies) combined.

---

[137] Michael Golebiewski and Danah Boyd, "Data Voids: Where Missing Data Can Easily Be Exploited," *Data & Society*, May 2018, pp. 3, 8.

[138] Interview with tech industry representative, San Francisco, Calif., February 6, 2019.

[139] Interview with tech company representative, San Francisco, Calif., April 10, 2019.

[140] Interview with tech company representative, San Francisco, Calif., February 6, 2019.

[141] Interview with tech industry representative, New York, N.Y., February 5, 2019.

Facebook and Twitter have expressed hesitancy in serving as "arbiters of truth" or determining what constitutes disinformation. Instead, they have opted to outsource these decisions to third-party groups and then enforce their own terms of service rather than any particular state interest.[142] Doing so provides a degree of distance from geopolitical debates and prevents these companies from being viewed simply as a tool of any particular state or viewpoint.[143]

While some degree of impartiality might be crucial to these companies' business interests, it also limits what the USG can expect from private industry in combating disinformation. Social media companies can be expected to curtail spam, fraud, threats of violence, or other criminal activity on their platforms.[144] To a lesser extent, these companies can be expected to clamp down on disinformation when there is significant outcry (as in the case of election meddling) or when there is clear harm to the public interest. For example, both Pinterest and Facebook tried to curtail material from anti-vaccination groups—a movement that contributed to the worst outbreak of measles in the United States in two decades.[145]

These companies might not clamp down on disinformation that harms DoD's interests, however, unless those actions also infringe on these companies' terms of service. Anti-American sentiment to U.S. overseas postures, for example, might not generate the same pressure for these companies to act as election interference would, especially if the sentiment is partially grounded in truth. Ultimately, the USG cannot simply outsource the disinformation fight to others; therefore, it must own the problem set.

---

[142] Brian Stelter, "Interview with Twitter CEO, Jack Dorsey," *Reliable Sources*, August 19, 2018; interview with tech industry representative, New York, N.Y., February 5, 2019; interview with tech company representative, San Francisco, Calif., February 6, 2019.

[143] Interview with think tank analyst, Washington, D.C., January 31, 2019.

[144] Interview with tech industry representative, San Francisco, Calif., February 26, 2019.

[145] Lena H. Sun, "Anti-Vaxxers Face Backlash as Measles Cases Surge," *Washington Post*, February 25, 2019; Laura Stampler, "How Pinterest Is Going Further than Facebook and Google to Quash Anti-Vaccination Misinformation," *Fortune*, February 20, 2019; Monika Bickert, "Combatting Vaccine Misinformation," Facebook Newsroom, March 7, 2019.

Figure 3.2
Twitter Requests for Removal of Content (Top Eight Countries)

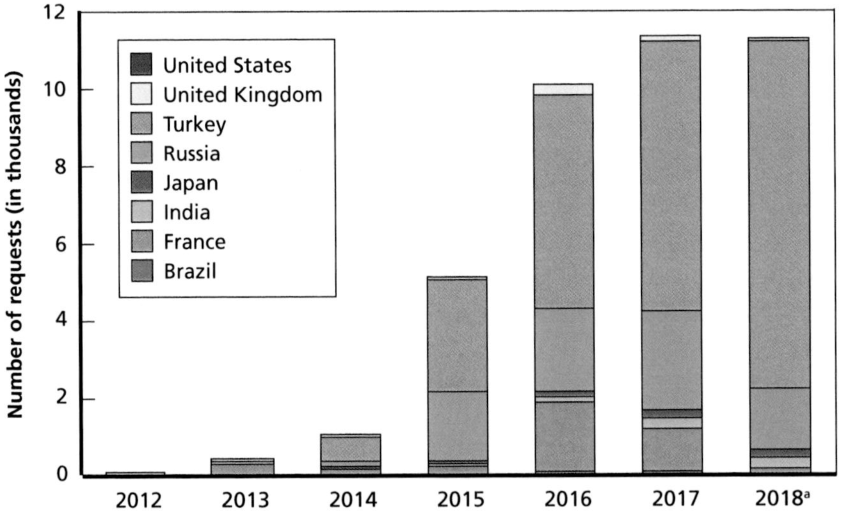

SOURCE: "Removal Requests," Twitter, undated.
[a] 2018 only includes information for part of the year.

Are Nongovernmental Organizations the Linchpin of This Fight?

Some of the most-important efforts to counter foreign disinformation campaigns come from outside governments and technology companies. Globally, there has been a groundswell of NGOs dedicated to detecting, analyzing, and combating disinformation over the past five years. Across the three regions studied in this project—Europe, the Indo-Pacific, and the United States—these civil society organizations are producing what is arguably some of the most-effective work to combat foreign disinformation, particularly in the consumption and awareness-raising categories of countermeasures (see Box 3.6).

In general, NGOs can provide information expertise that is not available in the public sector. For example, prior to the conflict, the Ukrainian government at large—and military in particular—realized that they lacked the skills and coordination to effectively counter Russian IO, so they partnered with the Ukraine Crisis Media Center (founded in March 2014) to tap into talent.[146] Some outside media advisors embedded in key part of ministries, including the Ministry of Defense and the General Staff.[147] The General Staff went one step further and built out an entire media team consisting of sociologists, psychologists, cameramen, and journalists to help monitor the information space for Russian disinformation and convey the Ukrainian military's story on both traditional and social media. [148]

Even when not providing direct support to governments in crisis, NGOs—particularly investigative journalism organizations—perform a critical role in debunking disinformation. The British-based online investigation outfit Bellingcat, for example, played a critical role in debunking Russian falsehoods about the downing of the Malaysian airliner MH-17 and unearthing Russia's role in the Skripal poisoning.[149] Across the world, much of the effort to combat mostly home-grown disinformation falls to private entities. For example, prior to the 2018 Okinawa public referendum, LINE worked to fact-check news.[150] Similar fact-checking was done by Okinawa's local daily publications before the September 2018 gubernatorial election.[151] During the 2019 Philippines elections, media outlets Rappler and VERA Files teamed up with nine other news organizations and

---

[146] Ukraine Crisis Media Center, "About Press Center," webpage, undated.

[147] Interview with media expert, Kyiv, Ukraine, March 8, 2019.

[148] Interview with media expert, Kyiv, Ukraine, March 8, 2019.

[149] For example, see Bellingcat Investigation Team, "JIT Indictments and Reactions: Analyzing New Evidence Linking Separatists and Russian Officials to MH17," July 17, 2019; Moritz Rakuszitzky, "Third Suspect in Skripal Poisoning Identified as Denis Sergeev, High-Ranking GRU Officer," Bellingcat, February 14, 2019.

[150] "(Recruiting at LINE@) Information and Opinions on Fake News in the Okinawa Public Referendum [【LINE@で募集中】沖┌県民投票のフェイクニュ┌ス情報┌意見>]," *Okinawa Times*, January 7, 2019.

[151] "Why Is Hate Speech and Fake News Against Okinawa Spreading? Is There Hope for the Internet Era? 'Fact Check' Discussion (2) [なぜ沖┌に┌するフェイク情報、ヘイト言┌が流れるのか?ネットの時代に希望はあるのか? 「ファクトチェック」座談┌【2】]," *Ryūkyū Shimpō*, May 24, 2019; "Okinawa Dailies Fact-Check, Debunk Rumors Spread During Gubernatorial Race," *Mainichi*, October 1, 2018.

Box 3.6
What Nongovernmental Organizations Are Doing

Production [prevent actors from producing or ordering production of content]	**Distribution** [restrict actors from distributing content]	**Consumption** [build audience resilience, lower susceptibility to content]
• N/A	• Some NGOs work with social media companies to remove or downgrade disinformation	• Strategic Communication Ukraine Crisis Media Center's efforts to improve Ukrainian government communication • Debunking efforts: Bellingcat, Rappler Vera File, StopFake, Atlantic Council's DFR Lab
Detection and Awareness-Raising		
• Bellingcat, Rappler Vera File, StopFake, and Atlantic Council's DFR Lab search for disinformation and attempt to publicize it		
Institution-Building		
• N/A		

NOTE: This list is illustrative rather than comprehensive.

three universities to launch a fact-checking website.[152] In Singapore, an alliance of regional media companies, including the *Straits Times*, has taken the initiative to raise awareness about fake news to help people become better consumers of online content.[153]

Even if governments can detect and debunk false claims themselves, going through outside fact-checking organizations provides two major benefits for the disinformation fight. In some cases, debunking of disinformation that is completed by a private organization can have greater international reach than efforts by governmental sources. For example, StopFake produces content in 11 languages: Bulgarian, Czech, Dutch, English, French, German, Italian, Polish, Romanian, Russian, and Spanish.[154] The organization now boasts podcasts, three television shows, and radio shows that are syndicated to Hromadske radio and are broadcast across the contact line in the Donbass. StopFake is also working with Radio Free Europe/Radio Liberty to broadcast across Crimea.[155]

Second and more importantly, information coming from an outside, independent organization might be viewed as more honest and credible than if the same message were to come from a government source. Certainly, from the technology sector's perspective, relying on these NGOs to label disinformation carries less reputational cost than doing this work in-house or relying on government judgements. As an interviewee from one of these fact-checking organizations put it,

> We don't get data from Facebook, but their funding gives us unrestricted funding to do the work we were doing, prior to them coming in and saying they have a big problem and need to spend money on it. Them partnering with us gives them more top cover

---

[152] Tsek.ph, "About Tsek.ph," webpage, undated.

[153] Shefali Rekhi, "ST to Share Insights from Fight Against Fake News," *Straits Times*, October 26, 2017.

[154] StopFake.org, "About Us," webpage, undated.

[155] Interview with journalist, Kyiv, Ukraine, March 5, 2019.

> to take more action on their platform. . . . That's a trade-off I'm willing to take for the foreseeable future.[156]

In other words, fact-checking organizations provide a useful political compromise. They can provide the missing connective tissue in the disinformation fight between a wary tech sector and a government seeking to counter foreign disinformation.

Although these outside fact-checkers play an important role in the counter-disinformation fight, the joint force cannot simply delegate this task to them altogether. First, some of these organizations have limited bandwidth. For example, despite the active Chinese disinformation campaign against Taiwan, Taiwan's independent fact-checker operations are still rather small: CoFacts has roughly ten part-time volunteer editors; Taiwan Fact Check Center has four full-time staff.[157] Second, these organizations need to be independent of government control to perform their roles as honest brokers of information—which means that their interests will not be wholly aligned with the USG or joint force. Therefore, the United States needs to retain its own capability in this field.

Conclusion: The Unsolved Challenge

The common thread across all the various government, allied, and private actors involved in the the counter-disinformation effort is that no one believes the problems is fully solved. Some measures have improved abilities to detect and then debunk disinformation. Others have arguably made society more resilient to disinformation's effects. Still others have aimed to deter disinformation altogether. Most experts in this area said that the disinformation problem will continue to grow in years to come. Therefore, the challenge for USAF and the joint force might not be how to stop foreign disinformation efforts; rather, it will be how to best mitigate the effects of these efforts. We tackle this topic in the next chapter.

---

[156] Interview with fact-checking organization, Washington, D.C., February 26, 2019.

[157] Interviews with social media experts, Taipei, Taiwan, January 2019.

CHAPTER FOUR

Conclusions and Recommendations

Our study resulted in mixed findings. As we discussed in Chapter Two, U.S. adversaries' use of disinformation on social media as a tool of statecraft might be more nuanced than is commonly portrayed. As we discussed in Chapter Three, the United States, other states, and industry and civil society groups are all struggling with how to respond effectively to these campaigns. Disinformation campaigns on social media will likely increase in the coming years, so we have developed a series of recommendations—across the entire framework of potential responses—for how USAF, the joint force, and USG at large can prepare to combat disinformation campaigns on social media in the coming years (see Box 4.1).[1]

Recommendations for AFSOC

The challenge of combating disinformation campaigns on social media goes well beyond any single service, let alone command.

[1] For an alternative comprehensive take on how the United States should respond to political warfare—a category that includes disinformation campaigns—see Ross Babbage, Thomas G. Mahnken, and Gillian Evans, *Winning Without Fighting: Chinese and Russian Political Warfare Campaigns and How The West Can Prevail*, Vol. 1, Washington, D.C.: Center for Strategic and Budgetary Assessments, 2019; Ross Babbage, Mike Winnerstig, Whitney McNamara, Grant Newsham, Anne-Marie Brady, Bob Lowry, and Nadège Rolland, *Winning Without Fighting: Chinese and Russian Political Warfare Campaigns and How The West Can Prevail*, Vol. 2, Washington, D.C.: Center for Strategic and Budgetary Assessments, 2019. Many recommendations of the Babbage et al. studies neatly align with ours.

Box 4.1
Recommendations

Production [prevent actors from producing or ordering production of content]	**Distribution** [restrict actors from distributing content]	**Consumption** [build audience resilience, lower susceptibility to content]
• AFSOC/USAF: Weigh Commando-Solo deployments in adversary's "near abroad" • Joint force: Increase transparency and enforce message discipline • USG: Focus offensive influence efforts on truthful information and weigh the risks carefully	• USG: Leverage industry but do not outsource the counter-disinformation fight • USG: Avoid bans on social networks	• Joint force: Train for disinformation, focus on key demographics, and minimize bans on smartphones and social media use • USG: Leverage civil society groups
Detection and Awareness-Raising • Joint force: Know the information environment and look beyond U.S. platforms • USG: Balance counter-disinformation with a commitment to freedom of speech		
Institution-Building • AFSOC/USAF: Expand IO capabilities and focus on more than operational security • Joint force: Know the information environment and look beyond U.S. platforms • Joint force: Conduct a DoD-wide review of the structure and authorities of the IO forces • USG: Publish a counter-disinformation strategy		

Nonetheless, we have two AFSOC-specific recommendations and implications.

Weigh Commando-Solo Deployments in Adversary's Near Abroad

China, Russia, North Korea, and Iran have embraced disinformation on social media as a weapon of conflict, driven in part by their own deep-seated anxieties about the possible impact of the internet and social media on domestic stability. All these countries are worried to varying degrees that the West will mount disinformation campaigns against them—particularly on social media—and spark unrest. Russian writings often exaggerate the West's actual capabilities in this arena and portray Russia's own disinformation efforts as merely trying to keep up with Western dominance.[2]

Consequently, U.S. adversaries might be hypersensitive to deployments of psychological operations assets—including EC-130J Commando-Solo—particularly in their immediate surroundings. For example, a Russian author in 1999 said that NATO sought to contest the "information space" through new technology such as the Commando-Solo airborne television and radio broadcasting platform, adding that this form of conflict presented a dire threat to Russia's security.[3] Other Russian military officers expressed similar concerns, particularly about Commando-Solo.[4]

A possible response to foreign disinformation efforts would be to deploy Commando-Solo or other IO assets to Russia or other adversar-

---

[2] Nikolai Borskiy, "Main Directions for Ensuring Information Security in the Activities of Troops (Forces) [Основные направления обеспечения информационной безопасности в деятельности войск (сил)]," *Orienteer [Ориентир]*, No. 11, November 2001; V. Belous, "Weapons of the 21st Century [Оружия XXI века]," *International Life [Международная Жизнь]*, No. 2, 2009.

[3] Yevgeniy Georgievich Zushin, "Power Has No Equal in Strength [Власть, не имеющая равных по силе воздействия]," *Independent Military Review [Независимое военное обозрение]*, No. 16, April 30, 1999; "Military Sites at the Festival for Author's Song [Военные площадки на фестивалях авторской песни]," Desantura.Ru, undated.

[4] Gennadiy Zhilin, "Information-Psychological Weapons: Yesterday and Today [Информационно-психологическое оружие: Вчера и сегодня]," *Soldier of the Fatherland [Солдат Отечества]*, No. 57, July 21, 2004.

ies near abroad. It is unclear whether doing so would serve as a deterrent or a provocation, but such a move would almost certainly attract an adversary's attention, and the risks and benefits of such deployment should be weighed carefully.

Expand USAF Information Operations Capabilities and Responsibilities

The preliminary data gathered for this study suggest that USAF will need to regrow its IO capability as the disinformation campaigns on social media increasingly become a staple of competition—and, perhaps, conflict. In this respect, the USAF decision to create a separate career field in 14F in 2016 and an IO school in 2018 was a step in the right direction.[5] Similarly, there are reports that USAF's 24th Special Operations Wing might develop a more robust social media capability. If these reports are true, this action is also a positive step.[6]

Still, as mentioned in Chapter Three, USAF IO needs to expand its focus and its capacity. During the height of Iraq and Afghanistan wars, much of IO concentrated on operational security concerns, watching for tail spotters and accidental leaks of sensitive data. In this new age of great-power competition, USAF information officers—like their counterparts throughout the joint force—need to proactively detect—and, if need be, counter—disinformation campaigns, particularly online.

Recommendations for USAF and the Joint Force

Aside from the AFSOC specific recommendations, there are several recommendations that more broadly apply to USAF and the joint force at large.

---

[5] Air Force Public Affairs, "AF Officials Announce Creation of Info Ops Tech School," March 5, 2018.

[6] Diana Stancy Correll, "Air Force's 24th Special Operations Wing Signals It Wants to Expand Social Media Operations," *Air Force Times*, August 20, 2019.

Know the Information Environment and Look Beyond U.S. Platforms

Although American social media platforms, such as Facebook, dominate the global market share, they are not always at the forefront of the information fight as we discussed in Chapter Two. Smaller platforms, such as Reddit or 4chan, might lack the resources (and, in some cases, the will) to combat disinformation. Locally popular platforms, such as PTT or dcinside, might be a better medium for an adversary to reach its target population.

For a joint force tasked with executing missions globally, the counterdisinformation problem is compounded by variations in the information environment, which force information operators to master different platforms, languages, and cultures. WebOps will face challenges as it transitions from a CENTCOM-focused asset to a part of the Joint MISO Warfare Center with a global mandate under Special Operations Command. Having much of the joint force's social media operations centralized in one organization could facilitate coordination, but the need for tailored language and cultural understanding—and a grasp of the local social media platforms—could limit some of the potential economies of scale. It will increase the burden—particularly on the U.S. Army, as WebOps' primary force provider—to provide these more-tailored forces and could drive demand for the other services to supply more of this capability.

For operational forces that regularly deploy to multiple regions of the world, the need to tailor IO capabilities to the local environment might cause even greater concerns. A force that lacks local expertise for a given information environment might need to hire outside contractors or partner with local organizations. Delegating to outsiders, however, poses other problems, from finding the right people to ensuring that those contracted people can be trusted with sensitive (and, at times, subjective) tasking. The joint force cannot necessarily predict with certainty where the next conflict will occur, but it should anticipate problems with delegating some of this mission to contractors and build in organizational structures that will be needed to deal with these issues when they arise.

Train for Disinformation, Focus on Key Demographics, and Minimize Widespread Bans on Smart Phone and Social Media Use

The joint force has made significant strides toward incorporating counter-disinformation efforts on social media into their regimen. One way has been to simulate the social media environment through such programs as ION. Another has been to make real-world information environment monitoring a part of regular training exercises, such as Trident Juncture. There are pros and cons to both approaches, but detecting and responding to disinformation needs to be built into training programs—from the unit level down to the individual service member level—particularly as disinformation on social media becomes a staple of great-power competition.

This expansion of disinformation training will push the joint force to reach new audiences, possibly in uncomfortable ways. U.S. adversaries do not target all service members equally; they might not even narrow their focus to just service members. China tends to base its disinformation campaigns on ethnic lines, placing Chinese- and Taiwanese-Americans at greater risk of attack.[7] Russia targets military family members. In 2015 and 2016, several military spouses—many of whom led military family support groups or wrote about military family matters—received death threats on Facebook and Twitter from a group claiming to be "Cyber Caliphate" but actually tied to Russian intelligence, possibly in a bid to deflect attention from Russia's actions in Ukraine and encourage support for Russia's action to fight the Islamic states in the Syrian conflict.[8] Both Russia and China could launch disinformation campaigns that do not target U.S. audiences at all but still hamper the joint force's ability to operate by targeting local communities where the joint force bases overseas. Properly training for disinformation campaigns requires reaching out to new audiences (such as family members and base communities) and tailoring modules for specific at-risk groups.

---

[7] Interview with Chinese disinformation expert, Taipei, Taiwan, January 2019.

[8] Cyber Caliphate being a Russian operation became "the consensus view among Western intelligence services." John R. Shindler, "False Flags: The Kremlin's Hidden Cyber Hand," *The Observer*, June 18, 2016; Raphael Satter, "Russian Hackers Posed as IS to Threaten Military Wives," Associated Press, May 8, 2018.

Training all service members and family members to recognize disinformation campaigns is a herculean task. A simpler solution would be to ban the use of smartphones and social media during periods of conflict. However, such a policy could prove impractical. As Ukraine's experience during the height of the conflict in the Donbass demonstrated, smartphone bans, even for front-line soldiers, proved difficult to enforce because bored soldiers found ways to smuggle phones to the front.[9] Theoretically, the joint force could pursue limited technological solutions (e.g., jamming reception or blocking access to certain sites) or selectively enforce bans for certain key units for limited periods of time, but improving force resiliency against disinformation will still depend mostly on training.

Increase Transparency and Enforce Message Discipline

Another part of increasing the joint force's resiliency to disinformation campaigns is denying would-be adversaries the opportunities to launch those campaigns in the first place. As we discussed in Chapter Two, many disinformation campaigns are at least partially true. The joint force cannot prevent every conspiracy theory or rumor from taking hold, but emphasizing transparency to the extent possible would reduce the efficacy of those disinformation efforts.

As mentioned in Chapters Two and Three and explored in the greater detail in the accompanying volume *Russian Disinformation Efforts on Social Media*,[10] a cornerstone of Russia's successful disinformation efforts in Ukraine was a series of unforced errors by the Ukrainian government. Poor message discipline—conflicting narratives about the war in the Donbass—allowed Russia to prey on the Ukrainian public's fears and to portray the military situation as far more dire than it actually was.

The joint force is unlikely to face quite the same situation that Ukraine did from 2014 to 2015, but the basic lesson of clear, consistent, and ample communication remains applicable for two reasons. First, the potential for different units to give conflicting narratives is

---

[9] Interview with midgrade Ukrainian military officers, Kyiv, Ukraine, March 7, 2019.

[10] Treyger, Cheravitch, and Cohen, forthcoming.

real given the size of the joint force, the numerous potential stovepipes across different units, and the different services for information. If the joint force is not careful, future adversaries could exploit any misinformation to their advantage. Second, even bracketing the misinformation problem, sometimes the joint force functionally cedes the information ground altogether. By overclassifying material—especially material that will likely become public anyhow—the joint force hamstrings its own ability to define the narrative surrounding its actions and creates the opportunity for adversarial disinformation.

Conduct a DoD-Wide Review of the Information Operations Force

Finally, the joint force is probably overdue for a servicewide and DoD-wide senior-level review of the personnel, structure, and authorities pertaining to the IO force.[11] As noted in Chapter Three, the joint force's efforts to man, train, and equip operations on social media remain ad hoc and service-specific, and the authorities involved are in many cases still legacies of an era before social media and the global media environment.[12] Given the centrality of IO to the National Defense Strategy's focus on competition, DoD should undertake a comprehensive review and update IO with an eye toward great-power competition in the digital age. Such a review should look specifically at the legal authorities involved in conducting operations on social media, the equipment used across the joint force, and the different organizational structures for fielding IO forces and active-reserve component mix.[13]

---

[11] In some ways, this recommendation mirrors an inspector general finding in Michael J. Roark, *(U) Army Contracting Command-Redstone and Space and Missile Defense: Command Need to Improve Contract Oversight for the Web-Based Military Information Support Operations Contract*, January 18, 2017, Released by Department of Defense Office of the Inspector General, Freedom of Information Agreement request DoDOIG-2017-000246.

[12] For a detailed analysis of challenges with authorities, personnel and structure in the information force, see Cohen et al., 2021; William Marcellino, Meagan L. Smith, Christopher Paul, and Lauren Skrabala, *Monitoring Social Media: Lessons for Future Department of Defense Social Media Analysis in Support of Information Operations*, Santa Monica, Calif.: RAND Corporation, RR-1742-OSD, 2017.

[13] Marcellino et al., 2017, provides a similar recommendation.

Recommendations for the U.S. Government

The terms "whole of government" and "whole of society" have almost become clichés, but much of the response in countering disinformation on social media does fall outside DoD's lane, and the effort will require cooperation with other government agencies, industry, and civil society groups. We choose to highlight six possible reforms.

Publish a Counter-Disinformation Strategy

As described in Chapter Three, the USG has several initiatives across multiple agencies touching different aspects of the counter-disinformation fight. But there is no coordination among these various efforts, and there is no single champion with the power political to arbitrate between agencies' bureaucratic interests and oversee the broader effort. Determining whether the United States needs to bring back the USIA in some form (or if it can leave these duties with a more robust version of GEC) was mostly beyond the scope of this project, but the USG can add some clarity by producing a counter-disinformation strategy, detailing how various elements will work together in a coherent fashion. Such a strategy should be openly published; doing so would prevent the strategy from becoming the target of disinformation, and it would serve as an invitation to external stakeholders—partners and allies, industry, and civil society groups—to participate in the broader counter-disinformation effort.

Leverage Industry but Do Not Outsource the Counter-Disinformation Fight

From the U.S. policy standpoint, one of the larger questions is how much the USG needs to be involved in fighting disinformation campaigns on social media versus the extent to which it can delegate this responsibility to industry and let social media companies police themselves. In this respect, this study generated mixed findings.

Social media companies have both the motivation and means to counter disinformation on their platforms better than the USG can. As mentioned in Chapter Three, larger U.S. companies—such as Facebook—have a vested interest in combating disinformation on their platforms, if only to avoid negative publicity that might drive away consumers. Social media companies also are arguably best positioned to verify the identities of users on their platforms and to detect and curb distribution of disinformation. Judging from interviews with company officials and with outside observers in the United States and around the world, many larger companies have taken concrete actions, particularly in the wake of the 2016 election interference, to crack down on disinformation.

That said, the USG cannot simply delegate counter-disinformation efforts to the private sector. Facebook, Twitter, and YouTube might have a stake in countering disinformation, but the same is not necessarily true for foreign or smaller platforms. Some of those business models revolve around anonymity and lack of restrictions on content, both of which allow disinformation campaigns to survive and thrive. It is also important to recognize that corporate and USG equities do not necessarily align even with larger U.S. platforms. Tech companies are for-profit entities with a global consumer base, after all. Countering disinformation campaigns around high-profile events, such as elections, might serve these companies' bottom lines, but the same might not be true of all cases that the USG cares about. Consequently, the USG needs to leverage industry but not outsource the counter-disinformation fight.

Leverage Civil Society Groups

Civil society groups play an increasingly important—but easy to overlook—role in the counter-disinformation fight. These groups lack the resources of government or industry, but outside fact-checking organizations can serve as impartial arbiters of disinformation and act as useful go-betweens for governments that want to combat disinformation and tech sectors that want to avoid the appearance of being too close to any specific state interests.

For the USG, the policy conundrum becomes how to leverage civil society groups in the counter-disinformation fight without jeopardizing their independence, which is the cornerstone of their legitimacy. One way to strike this balance would be to expand grant programs, such as those given by National Endowment for Democracy for "professional training in news gathering and reporting; assistance to journalistic associations and other groups dedicated to promoting and defending freedom of the press; and aid to print and electronic media that serve as forums for free discussion and the advancement of democratic ideas."[14] The strategic assumption supporting this approach is that disinformation campaigns would be curtailed in speed and scope by more investigative journalists chasing down falsehoods, rapidly and thoroughly debunking the material, and widely disseminating their findings.

Avoid Bans on Social Networks

As detailed in Chapter Three, the USG, its allies and partners, and states around the world have experimented with a variety of counter-disinformation efforts, such as providing media literacy programs, debunking false or misleading claims, and threatening prosecution and sanctions in retaliation for disinformation. In most cases, these efforts produce ambiguous results, with little comprehensive concrete evidence to prove their efficacy.

The possible exception to this rule is Ukraine's ban on VK, which produced measurable effects, although not necessarily positive ones. The ban arguably holds the title of being the single most forceful state action globally to counter disinformation. It is an open question whether the United States or any other liberal democracy not facing the same existential threat that Ukraine did could ever implement use of such a blunt tool. Even if it were possible, Ukraine's experience suggests that banning entire social networks does not solve the disinformation problem; instead, it creates significant second-order effects. As detailed in Chapter Three, the ban did not solve the Russian disinformation problem in Ukraine—and very well might have made VK into a more ideological platform with a younger audience than before.

---

[14] National Endowment for Democracy, "Founding Statement of Principles and Objectives, 1984," webpage, undated.

Balance Counter-Disinformation with a Commitment to Freedom of Speech

Particularly in the wake of the U.S. election interference in 2016, public debate has largely focused on how disinformation campaigns on social media can undermine democratic processes. Often overlooked in this debate is how counter-disinformation measures—if ham-fistedly applied—can have some of the same negative effects. Many of the more draconian distribution-focused countermeasures—blocking certain posts, certain actors, and even certain networks—raise significant concerns about freedom of speech and freedom of the press, especially because what qualifies as disinformation (as opposed to legitimate, albeit extreme, views) can be a subjective judgment call rather than a fact.

USG policy, therefore, must seek to counter disinformation without infringing on freedom of speech. This is true not only for U.S. domestic policy but also—perhaps even more so—for U.S. foreign policy. In newly emerging democracies and semiauthoritarian regimes, such objectives as countering disinformation can provide political cover for leaders who want to censor domestic opposition, stifle freedom of expression, and thus endanger the very institutions of democracy that counter-disinformation aims to protect.

Focus Offensive Influence Efforts on Truthful Information and Weigh the Risks Carefully

Finally, there is the question of whether the United States should develop its own offensive capability to conduct disinformation campaigns abroad as a potential deterrent to foreign campaigns targeting the United States, and its allies and partners. Two of the supporting volumes in this series discuss what offensive campaigns against Russia and China might look like and weigh the risks and benefits of doing so.

Ultimately, although there might be specific circumstances under which such efforts are warranted, all three analyses recommend caution in this area. First, there are practicalities: All these adversaries are authoritarian countries with controlled media spaces; this poses an obstacle to an offensive information effort. Second, given the deep-seated anxieties of these regimes regarding domestic unrest, simply developing U.S. capabilities to conduct large IO could be viewed as escalatory and detrimental to strategic stability. Third, any U.S. disinformation campaigns risk tainting both the reputation of the United States and the organically produced pro-democracy movements in these countries.[15] Disinformation campaigns have dubious strategic payoffs to begin with, so the United States should carefully study the costs and benefits before engaging in such a campaign.

A more promising offensive deterrent might be to develop information campaigns centered on truthful information. U.S. adversaries have genuine weaknesses—from rampant corruption to abuses of power—that could be exploited by U.S. IO without needing to manufacture false or misleading information. Such efforts would reduce some of the reputational and legitimacy dangers posed by information warfare. That said, even IO based solely on factual evidence would face practical obstacles and incur some of the same escalatory risks, so any such actions should still be pursued with caution.

Final Thoughts: No Magic Weapon (as of Now); No Magic Panacea

As in other domains, the United States today is in an arms race of sorts in the war of disinformation on social media. As mentioned in Chapter Two, social media itself is still relatively new and state-conducted disinformation campaigns on social media are even newer. Although disinformation campaigns on social media have not yet shown that they can shift strongly held prior beliefs wholesale, they have been proven to have localized operational effects. Moreover, in the years ahead, adversaries will be able to field increasingly sophisticated forms of disinformation. Advances in digital technology will

---

[15] For example, both Russia and China have tried to cast the prodemocracy protestors in Moscow and Hong Khong, respectively, as operatives of the United States. Mike Eckel, "How Russian Officials Are Spinning the Moscow Protests as a Foreign Plot," Radio Free Europe/Radio Liberty, August 5, 2019; Emily Feng and Amy Cheng, "China State Media Present Their Own Version of Hong Kong Protests," NPR, August 12, 2019.

likely mean that *deep fakes*—"highly realistic and difficult-to-detect digital manipulations of audio or video"—will likely become readily available for adversaries to try to manipulate their targets.[16] If unmitigated, the seeming authenticity of these forms of disinformation might give these campaigns greater effectiveness than we do today.[17]

At the same time, however, as noted in Chapter Three, society's response to disinformation is also evolving. Particularly after the 2016 election interference, the technology companies increased the attention and resources devoted to these challenges. Civil society groups sprouted up all over the world to combat disinformation. And governments, too—including that of the United States—have acted. In sum, who will "win" the arms race for disinformation in the end remains an open question.

Even if all the recommendations were followed, the United States would still lack a silver bullet in the counter-disinformation fight. That is partly a function of this domain of conflict. Perhaps, one expert from Taiwan put it best when he argued that defending against social media disinformation should be treated akin to ballistic missile defense, with "multilayered defenses required—these should be whole of military, whole of government, and whole of society."[18] Although this expert was talking specifically about Taiwan, the same lesson applies to the United States. Any comprehensive solution addressing disinformation would require the work of multiple actors—both public and private—at every stage of the disinformation production cycle. Any effort would need to evolve as both the geopolitical picture and technology evolve.

---

[16] Chesney and Citron, 2019.

[17] Mazarr et al., 2019.

[18] Interview with Taiwan government official, Taipei, Taiwan, January 2019.

References

Ackerman, Robert, "The Navy Turns a Sharper Eye Toward Information Warfare," *The Cyber Edge*, February 1, 2019. As of July 9, 2019:
https://www.afcea.org/content/navy-turns-sharper-eye-toward-information-warfare

"Advancing the Final Victory with a Revolutionary Offensive Offensive: Speech by Dear Kim Jong-Un at the 8th Annual Military Conference of the Workers' Party of Korea [혁명적인 사상공세로 최후승리를 앞당겨나가자: 경애하는 김정은 동지께서 조선로동당 제8차 사상일군대회에서 하신 연설]," *Labor News* [*Rodong Sinmun,* 노동신문], 2014.

Air Force Public Affairs, "AF Officials Announce Creation of Info Ops Tech School," March 5, 2018. As of August 1, 2019:
https://www.af.mil/News/Article-Display/Article/1457978/af-officials-announce-creation-of-info-ops-tech-school/

Akhmadullin, Vladimir, "The Word, Equal to the Bomb [Слово, приравненное к бомбе]," *Independent Military Review [Независимое военное обозрение]*, No. 25, July 2, 1999.

Allen, Jonathan, "Tony Blinken's Star Turn," *Politico*, September 16, 2019. As of June 20, 2019:
https://www.politico.com/story/2013/09/tony-blinkens-star-turn-096847

Armstrong, Matthew, "No, We Do Not Need to Revive the U.S. Information Agency," War on the Rocks, November 12, 2015. As of July 8, 2019:
https://warontherocks.com/2015/11/no-we-do-not-need-to-revive-the-u-s-information-agency/

Associated Press, "Report: CENTCOM Botches Effort to Fight Online Recruiting by Islamic State," *Tampa Bay Times*, January 31, 2017. As of June 25, 2019:
https://www.tampabay.com/news/military/macdill/report-centcom-botches-effort-to-fight-online-recruiting-by-islamic-state/2311577

Babbage, Ross, Thomas G. Mahnken, and Gillian Evans, *Winning Without Fighting: Chinese and Russian Political Warfare Campaigns and How The West Can Prevail*, Vol. 1, Washington, D.C.: Center for Strategic and Budgetary Assessments, 2019. As of December 6, 2019: https://csbaonline.org/uploads/documents/Winning_Without_Fighting_Final.pdf

Babbage, Ross, Mike Winnerstig, Whitney McNamara, Grant Newsham, Anne-Marie Brady, Bob Lowry, and Nadège Rolland, *Winning Without Fighting: Chinese and Russian Political Warfare Campaigns and How The West Can Prevail*, Vol. 2, Washington, D.C.: Center for Strategic and Budgetary Assessments, 2019. As of December 6, 2019: https://csbaonline.org/uploads/documents/Winning_Without_Fighting_Annex_Final2.pdf

Barker, Tyson, "Germany Strengthens Its Cyber Defense: How It's Meeting the Russian Threat," *Foreign Affairs*, May 26, 2017. As of August 28, 2019: https://www.foreignaffairs.com/articles/germany/2017-05-26/germany-strengthens-its-cyber-defense

Barojan, Donara, "#PutinAtWar: Social Media Surge on Skripal," *Medium*, April 5, 2018a. As of August 20, 2019: https://medium.com/dfrlab/putinatwar-social-media-surge-on-skripal-b5132db6f439

———, "#PutinAtWar: Disinformation Targets Trident Juncture," *DFRLab* via StopFake.org, November 8, 2018b. As of August 20, 2019: https://www.stopfake.org/en/putinatwar-disinformation-targets-trident-juncture/

Beauchamp-Mustafaga, Nathan, and Michael Chase, *Borrowing a Boat Out to Sea: The Chinese Military's Use of Social Media for Influence Operations*, Washington, D.C.: John Hopkins School of Advanced International Studies, 2019.

Bellingcat Investigation Team, "JIT Indictments and Reactions: Analyzing New Evidence Linking Separatists and Russian Officials to MH17," July 17, 2019. As of July 31, 2019: https://www.bellingcat.com/news/uk-and-europe/2019/07/17/jit-indictments-and-reactions-analyzing-new-evidence-linking-separatists-and-russian-officials-to-mh17/

Belous, V., "Weapons of the 21st Century [Оружия XXI века]," *International Life [Международная Жизнь]*, No. 2, 2009, pp. 64–82.

Bentzen, Naja, "Foreign Influence Operations in the EU," European Parliamentary Research Service, July 2018. As of August 20, 2019: http://www.europarl.europa.eu/RegData/etudes/BRIE/2018/625123/EPRS_BRI(2018)625123_EN.pdf

Biały, Beata, "Social Media—From Social Exchange to Battlefield," *Cyber Defense Review*, Vol. 2, No. 2, Summer 2017.

Bickert, Monika, "Combating Vaccine Misinformation," Facebook Newsroom, March 7, 2019.

Bodine-Baron, Elizabeth, Todd C. Helmus, Andrew Radin, and Elina Treyger, *Countering Russian Social Media Influence*, Santa Monica, Calif.: RAND Corporation, RR-2740-RC, 2018. As of August 28, 2019:
https://www.rand.org/pubs/research_reports/RR2740.html

Borskiy, Nikolai, "Main Directions for Ensuring Information Security in the Activities of Troops (Forces) [Основные направления обеспечения информационной безопасности в деятельности войск (сил)]," *Orienteer [Ориентир]*, No. 11, November 2001.

Boyd, Danah M., and Nicole B. Ellison, "Social Network Sites: Definition, History, and Scholarship," *Journal of Computer-Mediated Communication*, No. 13, 2008, pp. 210–230.

Brantly, Aaron, and Liam Collins, "A Bear of a Problem: Russian Special Forces Perfecting Their Cyber Capabilities," Association of the United States Army, November 28, 2018. As of July 22, 2019:
https://www.ausa.org/articles/bear-problem-russian-special-forces-perfecting-their-cyber-capabilities

Brattberg, Erik, and Tim Maurer, *Russian Election Interference: Europe's Counter to Fake News and Cyber Attacks*, Washington, D.C.: Carnegie Endowment for International Peace, May 23, 2018. As of August 12, 2019:
https://carnegieendowment.org/2018/05/23/russian-election-interference-europe-s-counter-to-fake-news-and-cyber-attacks-pub-76435

Brustein, Joshua, "Facebook Grappling with Employee Anger over Moderator Conditions," Bloomberg, February 25, 2019.

Chairman of the Joint Chiefs of Staff, *Charter of the Joint Information Operations Warfare Center*, Washington, D.C., CJSCI Instruction 5125.01, Washington, D.C., September 30, 2015. As of August 13, 2019:
https://www.jcs.mil/Portals/36/Documents/Library/Instructions/CJCSI%205125.01%C2%A0.pdf?ver=2017-02-08-175018-130

Chakrabarti, Samidh, "Fight Election Interference in Real Time," Facebook Newsroom, October 18, 2018.

Chen Wei-han, "MND Plays Down China Aircraft Threat," *Liberty Times*, December 19, 2016. As of August 7, 2019:
https://news.ltn.com.tw/news/focus/breakingnews/1921370

Chesney, Robert, and Danielle Citron, "Deepfakes and the New Disinformation War: The Coming Age of Post-Truth Geopolitics," *Foreign Affairs*, January/February 2019. As of August 22, 2019:
https://www.foreignaffairs.com/articles/world/2018-12-11/deepfakes-and-new-disinformation-war

China Power, "What Does China Really Spend on Its Military?" Center for Strategic and International Studies, undated. As of July 31, 2019:
https://chinapower.csis.org/military-spending/

Chinese State Council Information Office, China's Military Strategy, via Xinhua, May 2015. As of July 31, 2019:
http://english.gov.cn/archive/white_paper/2015/05/27/content_281475115610833.htm

Chung Li-hua and William Hetherington, "China Targets Polls with Fake Accounts," *Taipei Times*, November 5, 2018. As of July 29, 2019:
http://www.taipeitimes.com/News/front/archives/2018/11/05/2003703618

Cohen, Raphael S., Alyssa Demus, Michael Schwille, Nathan Vest, *U.S. Efforts to Combat Foreign Disinformation on Social Media*, Santa Monica, Calif.: RAND Corporation, 2021, Not available to the general public.

Cohen, Raphael S., and Andrew Radin, *Russia's Hostile Measures in Europe: Understanding the Threat*, Santa Monica, Calif.: RAND Corporation, RR-1793-A, 2019. As of June 19, 2019:
https://www.rand.org/pubs/research_reports/RR1793.html

Cole, J. Michael, "Fake News at Work: President Tsai 'Persecutes Religion in Taiwan,'" *Taiwan Sentinel*, July 20, 2017. As of July 29, 2019:
https://sentinel.tw/fake-news-tsai-persecutes-religion/

Collins, Ben, "On Reddit, Russian Propagandists Try New Tricks," NBC News, September 25, 2018. As of June 24, 2019:
https://www.nbcnews.com/tech/tech-news/reddit-russian-propagandists-try-new-tricks-n913131

Collins, Liam, "Russia Gives Lessons in Electronic Warfare," Association of the United States Army, July 26, 2018. As of July 22, 2019:
https://www.ausa.org/articles/russia-gives-lessons-electronic-warfare

Confessore, Nicholas, Gabriel J. X. Dance, Richard Harris, and Mark Hansen, "The Follower Factory," *New York Times*, January 27, 2018. As of August 7, 2019:
https://www.nytimes.com/interactive/2018/01/27/technology/social-media-bots.html

Correll, Diana Stancy, "Air Force's 24th Special Operations Wing Signals It Wants to Expand Social Media Operations," *Air Force Times*, August 20, 2019. As of August 21, 2019:
https://www.airforcetimes.com/news/your-air-force/2019/08/20/air-forces-24th-special-operations-wing-signals-it-wants-to-expand-social-media-operations/?utm_source=Sailthru&utm_medium=email&utm_campaign=EBB%20 08.21.19&utm_term=Editorial%20-%20Early%20Bird%20Brief

Costello, Ronan, "Celebrating #EUMediaLiteracyWeek," Twitter Blog, March 20, 2019.

dcinside management, "Inquiry Regarding North Korea's Disinformation Campaign," email correspondence with authors, July 9, 2019.

Deevy, Clair, "Introducing We Think Digital: New Digital Literacy Resources to Reach 1 Million People in Asia Pacific by 2020," Facebook Newsroom, March 4, 2019.

Dek, Anton, Kateryna Kononova, and Tetiana Marchenko, "The Effects of Banning the Social Network VK in Ukraine," in *Responding to Cognitive Security Challenges*, Riga, Latvia: NATO Strategic Communications Centre of Excellence, January 2019, pp. 39–58. As of July 17, 2019:
https://www.stratcomcoe.org/responding-cognitive-security-challenges

Deputy Commandant for Information, "Brief: MAGTF Operations in the Information Environment (OIE)," April 9, 2019.

Deutsche Welle, "Russia's Information Warfare Targets German Soldiers in Lithuania," Atlantic Council webpage, February 24, 2017. As of August 20, 2019:
https://www.atlanticcouncil.org/blogs/natosource/russia-s-information-warfare-targets-german-soldiers-in-lithuania

Dizard, Wilson P., Jr., *Inventing Public Diplomacy: The Story of the U.S. Information Agency*, London: Lynne Rienner Publishers, 2004.

Dobbins, James, Howard J. Shatz, and Ali Wyne, *Russia Is a Rogue, Not a Peer; China Is a Peer, Not a Rogue: Different Challenges, Different Responses*, Santa Monica, Calif.: RAND Corporation, PE-310-A, 2019. As of June 19, 2019:
https://www.rand.org/pubs/perspectives/PE310.html

Dorell, Oren, "State Department Launches $40 Million Initiative to Counter Russia Election Meddling," *USA Today*, February 26, 2018. As of June 28, 2019:
https://www.usatoday.com/story/news/world/2018/02/26/state-dept-launch-new-effort-counter-russian-election-meddling/371906002/

Eckel, Mike, "How Russian Officials Are Spinning the Moscow Protests as a Foreign Plot," Radio Free Europe/Radio Liberty, August 5, 2019. As of August 22, 2019:
https://www.rferl.org/a/russia-moscow-protests-foreign-plot-intelligence-spin/30093494.html

Ellis, Emma Grey, "4chan Is Turning 15—and Remains the Internet's Teenager," *Wired*, June 1, 2018. As of March 20, 2019:
https://www.wired.com/story/4chan-soul-of-the-internet/

European Commission, "A Europe That Protects: The EU Steps Up Action Against Disinformation," press release, December 5, 2018. As of August 20, 2019:
https://europa.eu/rapid/press-release_IP-18-6647_en.htm

EU vs. Disinfo, "Year in Review: 1001 Messages of Pro-Kremlin Disinformation," webpage, January 3, 2019. As of August 20, 2019:
https://euvsdisinfo.eu/year-in-review-1001-messages-of-pro-kremlin-disinformation/

Everington, Keoni, "China's 'Troll Factory' Targeting Taiwan with Disinformation Prior to Election," *Taiwan News*, November 5, 2018. As of July 29, 2019: https://www.taiwannews.com.tw/en/news/3568146

Facebook Newsroom, "Company Info," webpage, undated. As of February 15, 2019: https://newsroom.fb.com/company-info/

FBI—*See* Federal Bureau of Investigation.

Federal Bureau of Investigation, "The FBI Launches a Combating Foreign Influence Webpage," August 30, 2018. As of August 11, 2019: https://www.fbi.gov/news/pressrel/press-releases/the-fbi-launches-a-combating-foreign-influence-webpage

Feng, Emily, and Amy Cheng, "China State Media Present Their Own Version of Hong Kong Protests," NPR, August 12, 2019. As of August 22, 2019: https://www.npr.org/2019/08/14/751039100/china-state-media-present-distorted-version-of-hong-kong-protests

FireEye Intelligence, "Suspected Iranian Influence Operation Leverages Network of Inauthentic News Sites & Social Media Targeting Audiences in U.S., UK, Latin America, Middle East," blog post, August 21, 2018. As of June 19, 2019: https://www.fireeye.com/blog/threat-research/2018/08/suspected-iranian-influence-operation.html

Galeotti, Mark, *Controlling Chaos: How Russia Manages Its Political War in Europe*, London: European Council of Foreign Relations, 2017.

Gallagher, Mary, and Blake Miller, "Can the Chinese Government Really Control the Internet? We Found Cracks in the Great Firewall," *Washington Post*, February 21, 2017. As of July 30, 2019: https://www.washingtonpost.com/news/monkey-cage/wp/2017/02/21/can-the-chinese-government-really-control-the-internet-we-found-cracks-in-the-great-firewall/?noredirect=on&utm_term=.44add65e2ce6

Garamone, Jim, "Global Integration Deserves More Attention, Selva Says," U.S. Department of Defense, June 18, 2019. As of July 8, 2019: https://www.defense.gov/explore/story/Article/1881159/global-integration-deserves-more-attention-selva-says/

GDPR—*See* General Data Protection Regulation.

General Data Protection Regulation, "GDPR FAQs," webpage, undated-a. As of August 20, 2019: https://gdpr.eu/faq/

———, "GDPR Key Changes," webpage, undated-b. As of August 20, 2019: https://gdpr.eu/what-is-gdpr/

"German Spy Agency Warns of Chinese LinkedIn Espionage," BBC News, December 10, 2017. As of July 30, 2019: https://www.bbc.com/news/world-europe-42304297

"Germany Fines Facebook for Underreporting Hate Speech Complaints," *DW*, July 2, 2019. As of September 9, 2019:
https://www.dw.com/en/germany-fines-facebook-for-underreporting-hate-speech-complaints/a-49447820

Gleicher, Nathaniel, "Removing Coordinated Inauthentic Behavior from China," Facebook Newsroom, August 19, 2019. As of August 28, 2019:
https://newsroom.fb.com/news/2019/08/removing-cib-china/

Golebiewski, Michael, and Danah Boyd, "Data Voids: Where Missing Data Can Easily Be Exploited," *Data & Society*, May 2018.

Government of Singapore, Factually website, undated. As of July 9, 2019:
https://www.gov.sg/factually

Government of South Korea, National Security Act, Korea Law Translation Center, 2011. As of August 26, 2019:
https://elaw.klri.re.kr/eng_service/lawView.do?hseq=26692&lang=ENG

Grossman, Derek, Nathan Beauchamp-Mustafaga, Logan Ma, and Michael S. Chase, *China's Long-Range Bomber Flights: Drivers and Implications*, Santa Monica, Calif.: RAND Corporation, RR-2567-AF, 2018. As of July 29, 2019:
https://www.rand.org/pubs/research_reports/RR2567.html

Grundy, Tom, "Did China's State-Run News Agency Purchase Twitter Followers?" *Hong Kong Free Press*, April 14, 2015. As of August 7, 2019:
https://www.hongkongfp.com/2015/04/14/did-chinas-state-run-news-agency-purchase-twitter-followers/

Haltiwanger, John, "Russia Committed Act of War With Election Interference, Nikki Haley Says," *Newsweek*, October 19, 2017. As of August 4, 2019:
https://www.newsweek.com/russia-committed-act-war-election-interference-nikki-haley-says-688518

Harold, Scott W., Nathan Beauchamp-Mustafaga, and Jeffrey W. Hornung, *Chinese Disinformation Efforts on Social Media*, Santa Monica, Calif.: RAND Corporation, RR-4373/3-AF, 2021. As of July 2021:
https://www.rand.org/pubs/research_reports/RR4373z3.html

Han, Kirsten, "Why Singapore's Moves to Curb 'Fake News' May Backfire," *Washington Post*, March 5, 2018. As of July 9, 2019:
https://www.washingtonpost.com/news/global-opinions/wp/2018/03/05/why-singapores-moves-to-curb-fake-news-may-backfire/?utm_term=.bafb92434223

Head, Jonathan, "Outlaw or Ignore? How Asia Is Fighting 'Fake News,'" BBC, April 4, 2018. As of August 20, 2019:
https://www.bbc.com/news/world-asia-43637744

Helmus, Todd C., Elizabeth Bodine-Baron, Andrew Radin, Madeline Magnuson, Joshua Mendelsohn, William Marcellino, Andriy Bega, and Zev Winkelman, *Russian Social Media Influence: Understanding Russian Propaganda in Eastern Europe*, Santa Monica, Calif.: RAND Corporation, RR-2237-OSD, 2018. As of August 16, 2019:
https://www.rand.org/pubs/research_reports/RR2237.html

House of Commons Digital, Culture, Media, and Sport Committee, *Disinformation and 'Fake News': Final Report*, London, February 18, 2019. As of June 20, 2019:
https://publications.parliament.uk/pa/cm201719/cmselect/cmcumeds/1791/1791.pdf

Howard, Philip N., Bharath Ganesh, and Dimitra Liotsiou, "The IRA, Social Media and Political Polarization in the United States, 2012–2018," Oxford Internet Institute, December 17, 2018. As of January 31, 2019:
https://comprop.oii.ox.ac.uk/research/ira-political-polarization/

Human Rights Watch, "South Korea: Cold War Relic Law Criminalizes Criticism," May 28, 2015. As of August 26, 2019:
https://www.hrw.org/news/2015/05/28/south-korea-cold-war-relic-law-criminalizes-criticism

Iasiello, Emilio J., "Russia's Improved Information Operations: From Georgia to Crimea," *Parameters*, Vol. 47, No. 2, 2017, pp. 51–63.

Insikt Group, *Beyond Hybrid War: How China Exploits Social Media to Sway American Opinion*, Boston, Mass.: Recorded Future, 2019.

Interfax-Ukraine, "Banned VK Social Network 4th in Internet Traffic in Ukraine in April," *Kyiv Post*, May 17, 2018. As of June 24, 2019:
https://www.kyivpost.com/business/banned-vk-social-network-fourth-internet-traffic-ukraine-april.html

International Institute for Strategic Studies, *Military Balance 2019*, London, February 2019.

Jacobs, Julia, "In Twitter Purge, Top Accounts Lose Millions or Followers," *New York Times*, July 12, 2018.

Jennings, Ralph, "In the Name of 'Fake News,' Asian Governments Tighten Control on Social Media," *Los Angeles Times*, February 8, 2019. As of July 9, 2019:
https://www.latimes.com/world/asia/la-fg-asia-fake-news-20190130-story.html

Kaiman, Jonathan, "Free Tibet Exposes Fake Twitter Accounts by China Propagandists," *The Guardian*, July 22, 2014. As of July 29, 2019:
https://www.theguardian.com/world/2014/jul/22/free-tibet-fake-twitter-accounts-china-propagandists

Kamalipour, Yahya R., and Nancy Snow, eds., *War, Media, and Propaganda: A Global Perspective*, Lanham, Md.: Rowman & Littlefield, 2004.

Kang Tae-Hwa [강태화], "North Korea 'Facebook Honey Trap' . . . Befriended Officials To Ask for Documents [북한 '페북 미인계'…공직자와 친구 맺어 자료 요구]," *JoongAng Ilbo* [중앙일보], 2016. As of August 5, 2019: https://news.joins.com/article/19711432

Kavanagh, Jennifer, and Michael D. Rich, *Truth Decay: An Initial Exploration of the Diminishing Role of Facts and Analysis in American Public Life*, Santa Monica, Calif.: RAND Corporation, RR-2314-RC, 2018. As of November 26, 2019: https://www.rand.org/pubs/research_reports/RR2314.html

Kim Jung-Woo [김정우], "North Korea's 'Internet Invasion' Is Flaring Up [최근 기승 부리는 북한의 '인터넷 남침]," *Chosun* [월간조선], 2013. As of June 2, 2019: http://monthly.chosun.com/client/news/viw.asp?ctcd=G&nNewsNumb=201308100009

King, Gary, Jennifer Pan, and Margaret E. Roberts, "How the Chinese Government Fabricates Social Media Posts for Strategic Distraction, Not Engaged Argument," *American Political Science Review*, Vol. 111, No. 3, 2017, pp. 484–501.

Knowledge@Wharton, "Why Social Media Is the New Weapon in Modern Warfare," January 17, 2019. As of June 20, 2019: https://knowledge.wharton.upenn.edu/article/singer-weaponization-social-media/

Ko Tin-yau, "How Fake News Led to Suicide of Taiwan Representative in Osaka," *EJInsight*, September 19, 2018. As of July 30, 2019: http://www.ejinsight.com/20180919-how-fake-news-led-to-suicide-of-taiwan-representative-in-osaka/

Kofman, Michael, Katya Migacheva, Brian Nichiporuk, Andrew Radin, Olesya Tkacheva, and Jenny Oberholtzer, *Lessons from Russia's Operations in Crimea and Eastern Ukraine*, Santa Monica, Calif.: RAND Corporation, RR-1498-A, 2017. As of November 26, 2019: https://www.rand.org/pubs/research_reports/RR1498.html

Kolesov, P., "Georgia's Information War Against South Ossetia and Abkhazia [Информационная война Грузии против Южной Осетии и Абхазии]," *Foreign Military Review [Зарубежное военное обозрение]*, No. 10, October 2008, pp. 18–21.

Kyodo, "Japan Plans to Take Steps Against 'Fake News' by June," *Japan Times*, January 14, 2019. As of August 20, 2019: https://www.japantimes.co.jp/news/2019/01/14/national/japan-plans-take-steps-fake-news-june/

Lapowsky, Issie, "NATO Group Catfished Soldiers to Prove a Point About Privacy," *Wired*, February 18, 2019. As of August 20, 2019: https://www.wired.com/story/nato-stratcom-catfished-soldiers-social-media/?verso=true

Latvian Public Broadcasting, "Latvia Shuts Down Sputnik Propaganda Website," March 29, 2018. As of August 20, 2019: https://eng.lsm.lv/article/society/society/latvia-shuts-down-sputnik-propaganda-website.a175627/

Layne, Nathan, "U.S. Imposes Fresh Russia Sanctions for Election Meddling," Reuters, December 19, 2018. As of September 10, 2019: https://www.reuters.com/article/us-usa-russia-sanctions-treasury/u-s-imposes-fresh-russia-sanctions-for-election-meddling-idUSKCN1OI27F

Lee Hyo-Suk [이효석], "Beautiful Facebook Friend May Be A Spy . . . North Korea's Cyber Terrorism Diversifies [미모의 페친, 알고보니 간첩일수도…북 사이버테러 다양화]," Yonhap News Agency [연합뉴스], 2016. As of August 5, 2019: https://www.yna.co.kr/view/AKR20160720105300004

Lee Kui-Won [이귀원], "North Korea Appropriated South Korean Resident Registration Number (RRN) to Spread Rumors About Sinking of the Cheonan [북, 주민번호 도용 '천안함 날조' 유포]," Yonhap News Agency [연합뉴스], 2010. As of June 19, 2019: https://www.yna.co.kr/view/AKR20100601133151043

Leung, Hillary, "Singapore Is the Latest Country to Propose Tough Legislation Against Fake News," *Time*, April 2, 2019. As of July 9, 2019: https://time.com/5562501/singapore-fake-news-law-freedom-speech/

Lima, Cristiano, "Facebook, Twitter Take Down Disinformation Campaigns Linked to Iran, Russia, Venezuela," *Politico*, January 31, 2019.

Lister, Tim, and Clare Sebastian, "Stoking Islamophobia and Secession in Texas—From an Office in Russia," CNN, October 6, 2017. As of August 20, 2019: https://www.cnn.com/2017/10/05/politics/heart-of-texas-russia-event/index.html

Lockie, Alex, "Dick Cheney: Russia Meddling in the US Election Could Be 'an Act of War,'" *Business Insider*, March 28, 2017. As of August 4, 2019: https://www.businessinsider.com/dick-cheney-russia-hack-no-question-act-of-war-2017-3

Losh, Jack, "Is Russia Killing Off Eastern Ukraine's Warlords?" *Foreign Policy*, October 25, 2016. As of July 17, 2019: https://foreignpolicy.com/2016/10/25/who-is-killing-eastern-ukraines-warlords-motorola-russia-putin/

Lu Hsin-hui, Hsieh Chia-chen, Yeh Tzu-kang, and Elizabeth Hsu, "Authorities Deny Rumor of Ban on Incense, Ghost Money Burning," *FocusTaiwan*, July 21, 2017. As of July 29, 2019: http://focustaiwan.tw/news/aipl/201707210016.aspx

Mackintosh, Eliza, "Finland Is Winning the War on Fake News. What It's Learned May Be Crucial to Western Democracy," CNN, May 2019. As of August 28, 2019: https://www.cnn.com/interactive/2019/05/europe/finland-fake-news-intl/

Manchester, Julia, "Dem Calls Russia Meddling 'Act of War,' Urges Cyber Attack on Moscow Banks," *The Hill*, July 17, 2018. As of August 13, 2019:
https://thehill.com/hilltv/
rising/397366-dem-rep-no-question-that-russia-hacking-effort-is-act-of-war

Marcellino, William, Meagan L. Smith, Christopher Paul, and Lauren Skrabala, *Monitoring Social Media: Lessons for Future Department of Defense Social Media Analysis in Support of Information Operations*, Santa Monica, Calif.: RAND Corporation, RR-1742-OSD, 2017. As of November 26, 2019:
https://www.rand.org/pubs/research_reports/RR1742.html

Mattis, Peter, "China's 'Three Warfares' in Perspective," War on the Rocks, January 30, 2018. As of July 29, 2019:
https://warontherocks.com/2018/01/chinas-three-warfares-perspective/

Mazarr, Michael J., Abigail Casey, Alyssa Demus, Scott W. Harold, Luke J. Matthews, Nathan Beauchamp-Mustafaga, and James Sladden, *Hostile Social Manipulation: Present Realities and Emerging Trends*, Santa Monica, Calif.: RAND Corporation, RR-2713-OSD, 2019. As of November 26, 2019:
https://www.rand.org/pubs/research_reports/RR2713.html

McLaughlin, Timothy, "Disinformation Is Spreading on WhatsApp in India—And It's Getting Dangerous," *The Atlantic*, September 5, 2018. As of June 24, 2019:
https://www.theatlantic.com/international/archive/2018/09/
fighting-whatsapp-disinformation-india-kerala-floods/569332/

Merriam-Webster, "Misinformation," webpage, undated. As of August 10, 2019:
https://www.merriam-webster.com/dictionary/misinformation

"Military Sites at the Festival for Author's Song [Военные площадки на фестивалях авторской песни]," Desantura.Ru, undated. As of August 12, 2019:
http://desantura.ru/forum/forum43/topic12205/

Minister of Information Policy of Ukraine, "About the Ministry," webpage, undated. As of August 12, 2019:
https://mip.gov.ua/en/content/pro-ministerstvo.html

Ministry of National Defense of the People's Republic of China, "Chinese Air Force Official: Weibo, WeChat Public Account Open [中国空军官方微博、微信公众号开通运行]," press release, November 10, 2015. As of August 22, 2019:
http://www.mod.gov.cn/edu/2015-11/10/content_4628124.htm

Monaghan, Jennifer, "Vkontakte Founder Says Sold Shares Due to FSB Pressure," *Moscow Times*, April 17, 2014. As of June 24, 2019:
https://www.themoscowtimes.com/2014/04/17/
vkontakte-founder-says-sold-shares-due-to-fsb-pressure-a34132

Mulvihill, Geoff, "US Official: Russia Using Social Media to Divide Americans," Associated Press, July 15, 2018. As of June 19, 2019:
https://www.apnews.com/2e11aadd40a349cdb020cb6fe25c1e30

Nakashima, Ellen, "Inside a Russian Disinformation Campaign in Ukraine in 2014," *Washington Post*, December 25, 2017. As of July 29, 2019: https://www.washingtonpost.com/world/national-security/inside-a-russian-disinformation-campaign-in-ukraine-in-2014/2017/12/25/f55b0408-e71d-11e7-ab50-621fe0588340_story.html?utm_term=.9c98a0eb6cfa

———, "U.S. Cyber Command Operation Disrupted Internet Access of Russian Troll Factory on the Day of 2018 Midterms," *Washington Post*, February 27, 2019. As of September 9, 2019: https://www.washingtonpost.com/world/national-security/us-cyber-command-operation-disrupted-internet-access-of-russian-troll-factory-on-day-of-2018-midterms/2019/02/26/1827fc9e-36d6-11e9-af5b-b51b7ff322e9_story.html

National Endowment for Democracy, "Founding Statement of Principles and Objectives, 1984," webpage, undated. As of August 21, 2019: https://www.ned.org/about/statement-of-principles-and-objectives/

National Library Board (Singapore), "Fact-Checking Using Multiple Sources," webpage, undated-a. As of July 9, 2019: http://www.nlb.gov.sg/sure/fact-checking-using-multiple-sources/

———, "S.U.R.E. Campaign," webpage, undated-b. As of July 9, 2019: https://sure.nlb.gov.sg/about-us/sure-campaign/

NATO—*See* North Atlantic Treaty Organization.

Naval Information Warfare Systems Command Public Affairs, "SPAWAR Changes Name to Naval Information Warfare Systems Command—Aligns Identity with Mission," June 3, 2019. As of July 9, 2019: https://www.navy.mil/submit/display.asp?story_id=109773

Nemr, Christina, and William Gangware, *Weapons of Mass Distraction: Foreign State-Sponsored Disinformation in the Digital Age*, Washington, D.C.: Park Advisors, March 2019. As of June 20, 2019: https://www.state.gov/wp-content/uploads/2019/05/Weapons-of-Mass-Distraction-Foreign-State-Sponsored-Disinformation-in-the-Digital-Age.pdf

Newsguard, "The Internet Trust Tool," webpage, undated. As of March 13, 2019: https://www.newsguardtech.com/#

Ng, Kelly, "The Big Read: In an Era of Fake News, the Truth May Not Always Be Out There," *Today*, June 2, 2017. As of August 20, 2019: https://www.todayonline.com/singapore/big-read-era-fake-news-truth-may-not-always-be-out-there

Nip, Joyce Y. M., and Chao Sun, "China's News Media Tweeting, Competing with US Sources," *Westminster Papers in Communication and Culture*, Vol. 13, No. 1, 2018, pp. 98–122.

North Atlantic Treaty Organization Strategic Communications Centre of Excellence, "About Strategic Communications," webpage, undated. As of August 20, 2019:
https://www.stratcomcoe.org/about-strategic-communications

Nosova, Viktoria, "Study: Russia Is Among the Top Five Countries with the Strongest Cyber Force," Vesti.ru, October 10, 2017. As of January 31, 2019:
http://hitech.vesti.ru/article/634311/

Obama, Barack, "Executive Order—Developing an Integrated Global Engagement Center to Support Government-Wide Counterterrorism Communications Activities Directed Abroad and Revoking Executive Order 13548," Washington, D.C.: White House, Executive Order 13721, March 14, 2016. As of July 8, 2019:
https://www.federalregister.gov/documents/2016/03/17/2016-06250/developing-an-integrated-global-engagement-center-to-support-government-wide-counterterrorism

Office of the Director of National Intelligence, *Assessing Russian Activities and Intentions in Recent U.S. Elections: The Analytic Process and Cyber Incident Attribution*, ICA 2017-01D, January 2017a. As of June 27, 2019:
https://www.dni.gov/files/documents/ICA_2017_01.pdf

———, "Background to *Assessing Russian Activities and Intentions in Recent US Elections:* The Analytic Process and Cyber Incident Attribution," Washington, D.C., January 6, 2017b. As of June 20, 2019:
https://www.dni.gov/files/documents/ICA_2017_01.pdf

"Okinawa Dailies Fact-Check, Debunk Rumors Spread During Gubernatorial Race," *Mainichi*, October 1, 2018. As of August 20, 2019:
https://mainichi.jp/english/articles/20181001/p2a/00m/0na/012000c

Olesen, Alexa, "Where Did Chinese State Media Get All Those Facebook Followers?" *Foreign Policy*, July 7, 2015. As of August 7, 2019:
http://foreignpolicy.com/2015/07/07/china-facebook-peoples-daily-media-soft-power/

Ongstad, Mike, "Social Media Is the Machine Gun of Modern Disinformation War," *The Hill*, October 26, 2018. As of June 20, 2019:
https://thehill.com/opinion/technology/413191-social-media-is-the-machine-gun-of-modern-disinformation-war

O'Sullivan, Donie, "How a Hacked American Nightclub Twitter Account Was Implicated in China's Information War," CNN, August 21, 2019. As of August 28, 2019:
https://www.cnn.com/2019/08/20/tech/twitter-china-us/index.html

Pan, Jason, "China Subverting Elections: Premier," *Taipei Times*, November 2, 2018. As of August 20, 2019:
http://www.taipeitimes.com/News/front/archives/2018/11/02/2003703459/1

Parliament (Singapore), Protection from Online Falsehoods and Manipulation, Bill Number 10/2019), April 1, 2019. As of July 9, 2019:
https://www.parliament.gov.sg/docs/default-source/default-document-library/protection-from-online-falsehoods-and-manipulation-bill10-2019.pdf

Paul, Christopher, and Miriam Matthews, *The Russian "Firehose of Falsehood" Propaganda Model: Why It Might Work and Options to Counter It*, Santa Monica, Calif.: RAND Corporation, PE-198-OSD, 2016. As of August 12, 2019:
https://www.rand.org/pubs/perspectives/PE198.html

People's Liberation Army Air Force, "Air Force Release [空军发布]," Weibo, undated. As of July 30, 2019:
https://m.weibo.cn/p/1005055707057078

Perano, Ursula, "YouTube Disables 210 Channels Linked to Hong Kong Influence Campaign," *Axios*, August 22, 2019. As of August 28, 2019:
https://www.axios.com/youtube-hong-kong-protests-facebook-twitter-fa4ff18b-e905-426e-92f1-57519b579923.html

PLA Air Force—*See* People's Liberation Army Air Force.

"PLA Air Force Releases Apparent H-6K Photographed with Taiwan's Jade Mountain [解放军空军发布疑似轰-6K与台湾玉山合影]," *Observer* [观察者], December 17, 2016. As of August 28, 2019:
http://www.guancha.cn/military-affairs/2016_12_17_384771.shtml

Pomerleau, Mark, "Why the Marine Corps Needed a New Deputy Commandant," *C4ISRNET*, December 5, 2017. As of June 25, 2019:
https://www.c4isrnet.com/it-networks/2017/12/05/why-the-marine-corps-needed-a-new-deputy-commandant/

———, "How the Navy Is Changing Its Thinking on Information Warfare," *C4ISRNET*, April 21, 2019. As of August 16, 2019:
https://www.c4isrnet.com/intel-geoint/isr/2019/04/22/how-the-navy-is-changing-its-thinking-on-information-warfare/

Popovych, Nataliia, and Oleksiy Makhuhin, "Countering Disinformation: Some Lessons Learnt by Ukraine Crisis Media Center," Ukraine Crisis Media Center, April 20, 2018. As of July 29, 2019:
http://uacrisis.org/66275-countering-disinformation-lessons-learnt

Priest, Dana, and Michael Birnbaum, "Europe Has Been Working to Expose Russian Meddling for Years," *Washington Post*, June 25, 2017. As of August 20, 2019:
https://www.washingtonpost.com/world/europe/europe-has-been-working-to-expose-russian-meddling-for-years/2017/06/25/e42dcece-4a09-11e7-9669-250d0b15f83b_story.html

Public Law 110-417, Subpart X, Duncan Hunter National Defense Authorization Act for Fiscal Year 2009, Sec. 1056, Prohibitions Relating to Propaganda, 110th Congress, October 14, 2009. As of July 8, 2019: https://www.govinfo.gov/content/pkg/PLAW-110publ417/pdf/PLAW-110publ417.pdf

Public Law 115-91, Div. A, National Defense Authorization Act for Fiscal Year 2018, December 2017. As of August 13, 2019: https://www.congress.gov/115/plaws/publ91/PLAW-115publ91.pdf

Rakuszitzky, Moritz, "Third Suspect in Skripal Poisoning Identified as Denis Sergeev, High-Ranking GRU Officer," Bellingcat, February 14, 2019. As of July 31, 2019: https://www.bellingcat.com/news/uk-and-europe/2019/02/14/third-suspect-in-skripal-poisoning-identified-as-denis-sergeev-high-ranking-gru-officer/

Reality Check Team, "Social Media: How Can Governments Regulate It?" BBC News, April 8, 2019. As of July 9, 2019: https://www.bbc.com/news/technology-47135058

"(Recruiting at LINE@) Information and Opinions on Fake News in the Okinawa Public Referendum [【LINE@で募集中】沖┌県民投票のフェイクニュ┌ス情報┌意見]," *Okinawa Times*, January 7, 2019.

Rekhi, Shefali, "ST to Share Insights from Fight Against Fake News," *Straits Times*, October 26, 2017. As of July 9, 2019: https://www.straitstimes.com/singapore/st-to-share-insights-from-fight-against-fake-news

"Removal Requests," Twitter, undated. As of August 12, 2019: https://transparency.twitter.com/en/removal-requests.html

Revelli, Alice, and Lee Foster, "Network of Social Media Accounts Impersonates U.S. Political Candidates, Leverages U.S. and Israeli Media in Support of Iranian Interests," FireEye Threat Research, blog post, May 28, 2019. As of August 20, 2019: https://www.fireeye.com/blog/threat-research/2019/05/social-media-network-impersonates-us-political-candidates-supports-iranian-interests.html

Roark, Michael J., *(U) Army Contracting Command-Redstone and Space and Missile Defense: Command Need to Improve Contract Oversight for the Web-Based Military Information Support Operations Contract*, January 18, 2017, Released by Department of Defense Office of the Inspector General, Freedom of Information Agreement request DoDOIG-2017-000246. As of September 16, 2019: https://media.defense.gov/2019/Mar/19/2002102869/-1/-1/1/DoDIG-2017-042%20(REDACTED).PDF

Robinson, Linda, Todd C. Helmus, Raphael S. Cohen, Alireza Nader, Andrew Radin, Madeline Magnuson, and Katya Migacheva, *Modern Political Warfare: Current Practices and Possible Responses*, Santa Monica, Calif.: RAND Corporation, RR-1772-A, 2018. As of July 8, 2019:
https://www.rand.org/pubs/research_reports/RR1772.html

Romm, Tony, "Facebook and Twitter Disable New Disinformation Campaign with Ties to Iran," *Washington Post*, May 29, 2019. As of June 19, 2019:
https://www.washingtonpost.com/technology/2019/05/28/facebook-twitter-disable-new-disinformation-campaign-with-ties-iran/?utm_term=.3e4f81cff5f0

Rosenberg, Matthew, Charlie Savage, and Michael Wines, "Russia Sees Midterm Elections as Chance to Sow Fresh Discord, Intelligence Chiefs Warn," *New York Times*, February 13, 2018. As of July 29, 2019:
https://www.nytimes.com/2018/02/13/us/politics/russia-sees-midterm-elections-as-chance-to-sow-fresh-discord-intelligence-chiefs-warn.html

Salvo, David, and Stephanie De Leon, "Russia's Efforts to Destabilize Bosnia and Herzegovina," Alliance for Securing Democracy, April 25, 2018. As of August 20, 2019:
https://securingdemocracy.gmfus.org/russias-efforts-to-destabilize-bosnia-and-herzegovina/

Satter, Raphael, "Russian Hackers Posed as IS to Threaten Military Wives," Associated Press, May 8, 2018. As of September 9, 2019:
https://www.apnews.com/4d174e45ef5843a0ba82e804f080988f

Schleifer, Theodore, and Deirdre Walsh, "McCain: Russian Cyberintrusions an 'Act of War,'" CNN, December 30, 2016. As of August 4, 2019:
https://www.cnn.com/2016/12/30/politics/mccain-cyber-hearing/index.html

Science and Technology Directorate, "Social Media Working Group for Emergency Services and Disaster Management," fact sheet, Washington, D.C.: U.S. Department of Homeland Security, November 22, 2017. As of August 13, 2019:
https://www.dhs.gov/sites/default/files/publications/SMWG_SMWG-Emergency-Services-Disaster-Management-FactSheet_171122-508.pdf

Schwirtz, Michael, "German Election Mystery: Why No Russian Meddling?" *New York Times*, September 21, 2017. As of August 28, 2019:
https://www.nytimes.com/2017/09/21/world/europe/german-election-russia.html

Shambaugh, David, "China's Soft-Power Push," *Foreign Affairs*, July 2015. As of July 29, 2019:
https://www.foreignaffairs.com/articles/china/2015-06-16/china-s-soft-power-push

Shin Bo-Young [신보영], "North Korea Appropriates South Korean RRN for Coordinated Propaganda [북, 남한 주민번호 도용 네티즌 조직적 선동]," *Culture Daily* [문화일보], 2010. As of July 17, 2019:
http://www.munhwa.com/news/view.html?no=20100601010301231160020

Shindler, John R., "False Flags: The Kremlin's Hidden Cyber Hand," *The Observer*, June 18, 2016. As of August 28, 2019:
https://observer.com/2016/06/false-flags-the-kremlins-hidden-cyber-hand/

Shultz, Richard H., and Roy Godson, *Dezinformatsiya: Active Measures in Soviet Strategy*, Washington, D.C.: Pergamon-Brassey, 1984.

Silverman, Craig, and Jane Lytvynenko, "Reddit Has Become a Battleground of Alleged Chinese Trolls," *BuzzFeed News*, March 14, 2019. As of August 28, 2019:
https://www.buzzfeednews.com/article/craigsilverman/reddit-coordinated-chinese-propaganda-trolls

Singapore government official, email with authors, June 21, 2019.

Snyder, Glenn Herald, *Deterrence and Defense: Toward a Theory of National Security*, Princeton, N.J.: Princeton University Press, 1961.

Social Media Working Group for Emergency Services and Disaster Management, *Countering False Information on Social Media in Disasters and Emergencies*, Washington, D.C.: Department of Homeland Security, March 2018. As of August 11, 2019:
https://www.dhs.gov/sites/default/files/publications/SMWG_Countering-False-Info-Social-Media-Disasters-Emergencies_Mar2018-508.pdf

Somer, Iryna, "Lithuanians Create Artificial Intelligence with Ability to Identify Fake News in 2 Minutes," *Kyiv Post*, September 21, 2018. As of February 4, 2019:
https://www.kyivpost.com/technology/lithuanian-creates-artificial-intelligence-with-ability-to-identify-fake-news-within-2-minutes.html

Stampler, Laura, "How Pinterest Is Going Further than Facebook and Google to Quash Anti-Vaccination Misinformation," *Fortune*, February 20, 2019.

Statista, "Number of Monthly Active WhatsApp Users Worldwide from April 2013 to December 2017 (in millions)," webpage, January 2018a. As of November 28, 2018:
https://www.statista.com/statistics/260819/number-of-monthly-active-whatsapp-users/

———, "Internet Penetration Rate in the Middle East Compared to the Global Internet Penetration Rate from 2009 to 2018," webpage, March 2018b. As of November 28, 2018:
https://www.statista.com/statistics/265171/comparison-of-global-and-middle-eastern-internet-penetration-rate/

———, "Number of Monthly Active Telegram Users Worldwide from March 2014 to March 2018 (in millions)," webpage, March 2018c. As of November 28, 2018:
https://www.statista.com/statistics/265171/comparison-of-global-and-middle-eastern-internet-penetration-rate/

———, "Number of Social Media Users Worldwide from 2010 to 2021 (in billions)," webpage, May 2018d. As of February 18, 2019: https://www.statista.com/statistics/278414/number-of-worldwide-social-network-users/

———, "Number of Monthly Active Instagram Users from January 2013 to June 2018 (in millions)," webpage, June 2018e. As of November 28, 2018: https://www.statista.com/statistics/253577/number-of-monthly-active-instagram-users/

———, "Number of Daily Active Snapchat Users from 1st Quarter 2014 to 3rd Quarter 2018 (in millions)," webpage, October 2018f. As of November 28, 2018: https://www.statista.com/statistics/545967/snapchat-app-dau/

———, "Number of Monthly Active Facebook Users Worldwide as of 3rd Quarter 2018 (in millions)," webpage, October 2018g. As of November 28, 2018: https://www.statista.com/statistics/264810/number-of-monthly-active-facebook-users-worldwide/

——— "Number of Monthly Active Twitter Users Worldwide from 1st Quarter 2010 to 4th Quarter 2018 (in millions)," webpage, February 2019. As of February 15, 2019: https://www.statista.com/statistics/282087/number-of-monthly-active-twitter-users/

Stelter, Brian, "Interview with Twitter CEO, Jack Dorsey," *Reliable Sources*, August 19, 2018. As of March 21, 2019: https://www.youtube.com/watch?v=Cm_lmWWKDug

Stephenson, Henry, "Navy Information Warfare: A Decade of Indulging a False Analogy," *Proceedings*, U.S. Naval Institute, Vol. 145, No. 1/1,391, January 2019. As of August 13, 2019: https://www.usni.org/magazines/proceedings/2019/january/navy-information-warfare-decade-indulging-false-analogy

StopFake.org, "About Us," webpage, undated. As of July 17, 2019: https://www.stopfake.org/en/about-us/

Strobel, Warren, "U.S. Losing 'Information War' to Russia, Other Rivals: Study," Reuters, March 25, 2015. As of July 31, 2019: https://www.reuters.com/article/us-usa-broadcasting idUSKBN0ML1MN20150325

Strong, Matthew, "Military Denies Yushan in China Bomber Picture: Peak Likely to Be Mount Beidawu in Southern Taiwan: Experts," *Taiwan News*, December 17, 2016. As of July 31, 2019: https://www.taiwannews.com.tw/en/news/3053731

Stubbs, Jack, and Christopher Bing, "Special Report: How Iran Spreads Disinformation Around the World," Reuters, November 30, 2018. As of June 19, 2019:
https://www.reuters.com/article/us-cyber-iran-specialreport/special-report-how-iran-spreads-disinformation-around-the-world-idUSKCN1NZ1FT

Sun, Lena H., "Anti-Vaxxers Face Backlash as Measles Cases Surge," *Washington Post*, February 25, 2019.

Swiecicki, Dominik, and Irene Christensson, "MSB Activities on Countering Information Influence Campaigns," Counter Influence Branch, Global Monitoring and Analysis Section, Swedish Civil Contingencies Agency, presentation at RAND Corporation in Santa Monica, Calif., November 29, 2018.

"Taiwan's Taoists Protest Against Curbs on Incense and Firecrackers," BBC News, July 23, 2017. As of July 29, 2019:
https://www.bbc.com/news/world-asia-40699113

Taylor, Guy, "State Department Global Engagement Center Targets Russian Propaganda, 'Deep Fakes,'" *Washington Times* via Associated Press, December 12, 2018. As of June 28, 2019:
https://www.apnews.com/9f7892a163582b5fd0297e2a81124c35

Theohary, Catherine A., *Information Warfare: Issues for Congress*, Washington, D.C.: Congressional Research Service, R45142, March 5, 2018. As of August 11, 2019:
https://fas.org/sgp/crs/natsec/R45142.pdf

"Third-Party Fact-Checking on Facebook," Facebook, undated. As of March 12, 2019:
https://www.facebook.com/journalismproject/programs/third-party-fact-checking

Toh, Michelle, "Google Says Singapore Risks Hurting Innovation with Fake News Law," CNN, May 9, 2019. As of July 9, 2019:
https://www.cnn.com/2019/05/09/tech/singapore-fake-news-law-tech/index.html

Tomlin, Gregory M., "#SocialMediaMatters: Lessons Learned from Exercise Trident Juncture," *Joint Force Quarterly*, No. 82, July 1, 2016. As of August 20, 2019:
https://ndupress.ndu.edu/JFQ/Joint-Force-Quarterly-82/Article/793264/socialmediamatters-lessons-learned-from-exercise-trident-juncture/

Treyger, Elina, Joe Cheravitch, and Raphael S. Cohen, *Russian Disinformation Efforts on Social Media*, Santa Monica, Calif.: RAND Corporation, RR-4373/2-AF, forthcoming.

Tsek.ph, "About Tsek.ph," webpage, undated. As of July 31, 2019:
https://tsek.ph/about

Tu, Aaron, and William Hetherington, "Defense Bureau to Tackle Propaganda from China," *Taipei Times*, March 4, 2019. As of July 31, 2019:
http://www.taipeitimes.com/News/taiwan/archives/2019/03/04/2003710821

Turovsky, Daniil, "'It's Our Time to Serve the Motherland': How Russia's War in Georgia Sparked Moscow's Modern-Day Recruitment of Criminal Hackers," *Meduza*, August 7, 2018. As of August 28, 2019:
https://meduza.io/en/feature/2018/08/07/it-s-our-time-to-serve-the-motherland

Twitter Safety, "Information Operations Directed at Hong Kong," Twitter Blog, August 19, 2019. As of August 28, 2019:
https://blog.twitter.com/en_us/topics/company/2019/information_operations_directed_at_Hong_Kong.html

"Ukraine Bans Its Top Social Networks Because They Are Russian," *The Economist*, May 19, 2017. As of June 24, 2019:
https://www.economist.com/europe/2017/05/19/ukraine-bans-its-top-social-networks-because-they-are-russian

Ukraine Crisis Media Center, "About Press Center," webpage, undated. As of July 17, 2019:
http://uacrisis.org/about

United Nations Educational, Scientific and Cultural Organization, *Journalism, 'Fake News' & Disinformation*, Paris, France, 2018. As of August 10, 2019:
https://en.unesco.org/sites/default/files/journalism_fake_news_disinformation_print_friendly_0.pdf

United States of America v. Elena Alekseevna Khusyaynova, U.S. District Court, Alexandria, Va., September 28, 2018.

United States of America v. Internet Research Agency et al., U.S. District Court, District of Columbia, February 16, 2018. As of March 26, 2020:
https://www.justice.gov/file/1035477/download

United States of America v. Park Jin Hyok, U.S. District Court for the Central District of California, June 8, 2018.

U.S. Advisory Commission on Public Diplomacy, *Consolidation of USIA into the State Department: An Assessment After One Year*, Washington, D.C.: U.S. Department of State, October 2000. As of August 11, 2019:
https://1997-2001.state.gov/policy/pdadcom/acpdreport.pdf

———, *2018 Comprehensive Annual Report on Public Diplomacy and International Broadcasting: Focus on FY2017 Budget Data*, Washington, D.C.: U.S. Department of State, 2018. As of July 8, 2019:
https://www.state.gov/wp-content/uploads/2019/05/2018-ACPD.pdf

USAF—*See* U.S. Air Force.

USAGM—*See* U.S. Agency for Global Media.

U.S. Agency for Global Media, "History," webpage, undated. As of June 24, 2019:
https://www.usagm.gov/who-we-are/history/

———, *FY 2018 Performance and Accountability Report*, Washington, D.C.: U.S. Broadcasting Board of Governors, November 2018.

U.S. Air Force, *Military Information Support Operations (MISO)*, Air Force Instruction 10-702, Washington, D.C., June 7, 2011. As of July 9, 2019:
https://fas.org/irp/DoDdir/usaf/afi10-702.pdf

U.S. Air Forces Cyber, *Gunslingers: A Brief History of the 67th Cyberspace Wing*, undated. As of August 13, 2019:
https://www.afcyber.af.mil/Portals/11/documents/
67%20CW%20Heritage%20Pamphlet%202017.pdf?ver=2017-11-21-145716-253

U.S. Army Cyber Command, "1st Information Operations Command (LAND)," webpage, undated-a. As of July 9, 2019:
https://www.arcyber.army.mil/Organization/1st-IO-Command/

———, "History," webpage, undated-b. As of July 9, 2019:
https://www.arcyber.army.mil/Organization/History/

U.S. Army Reserve, "U.S. Army Civil Affairs & Psychological Operations Command (Airborne)," webpage, undated. As of June 25, 2019:
https://www.usar.army.mil/Commands/Functional/USACAPOC/About-Us/

U.S. Army Training and Doctrine Command, "G-2 Operational Environment Center TRADOC G-2: Information Operations Network," webpage, undated. As of August 13, 2019:
https://ion.army.mil/

U.S. Code, Title 22, Section 2656, Management of Foreign Affairs.

U.S. Department of Defense, Summary of the National Defense Strategy of the United States of America: Sharpening the American Military's Competitive Edge, Washington, D.C., 2018. As of June 19, 2019:
https://www.defense.gov/Portals/1/Documents/pubs/
2018-National-Defense-Strategy-Summary.pdf

———, *Department of Defense Dictionary of Military and Associated Terms*, Washington, D.C., June 2019.

U.S. Department of State, "About Us—Global Engagement Center," webpage, undated-a. As of June 28, 2019:
https://www.state.gov/bureaus-offices/
under-secretary-for-public-diplomacy-and-public-affairs/global-engagement-center/

———, "Under Secretary for Public Diplomacy and Public Affairs: Our Mission," webpage, undated-b. As of July 8, 2019:
https://www.state.gov/bureaus-offices/
under-secretary-for-public-diplomacy-and-public-affairs/

U.S. Department of State, Office of the Spokesperson, "State-Defense Cooperation on Global Engagement Center Programs and Creation of the Information Access Fund to Counter State-Sponsored Disinformation," media note, February 26, 2018. As of July 8, 2019:
https://www.state.gov/state-defense-cooperation-on-global-engagement-center-programs-and-creation-of-the-information-access-fund-to-counter-state-sponsored-disinformation/

U.S. Marine Corps, "MCINCR—Marine Corps Base Quantico," webpage, undated. As of July 3, 2019:
https://www.quantico.marines.mil/Tenants/Marine-Corps-Information-Operations-Center/

USMC—*See* U.S. Marine Corps.

U.S. Securities and Exchange Commission, Form 10-K, Facebook, Inc., 2016. As of March 21, 2019:
https://www.sec.gov/Archives/edgar/data/1326801/000132680117000007/fb-12312016x10k.htm#s5611039F2AC75A779AB7D0EDD63A52CA

Vilmer, Jean-Baptiste Jeangène, Alexandre Escorcia, Marine Guillaume, and Janaina Herrera, *Information Manipulation: A Challenge for Our Democracies*, Paris: Policy Planning Staff (CAPS) of the Ministry for Europe and Foreign Affairs and the Institute for Strategic Research (IRSEM) of the Ministry for the Armed Forces, August 2018. As of August 20, 2019:
https://www.diplomatie.gouv.fr/IMG/pdf/information_manipulation_rvb_cle838736.pdf

Volz, Dustin, and David Ingram, "Facebook's Zuckerberg Unscathed by Congressional Grilling, Stock Rises," Reuters, April 11, 2018. As of July 9, 2019:
https://www.reuters.com/article/us-facebook-privacy-zuckerberg/facebooks-zuckerberg-unscathed-by-congressional-grilling-stock-rises-idUSKBN1HI1CJ

Wan, Cara, "No Need to Be Overly Worried About Fake News Laws, Says Ong Ye Kung," *Straits Times*, April 29, 2019. As of August 20, 2019:
https://www.straitstimes.com/singapore/no-need-to-be-overly-worried-about-fake-news-laws-says-ong

White House, National Security Strategy of the United States of America, Washington, D.C., December 2017. As of June 19, 2019:
https://www.whitehouse.gov/wp-content/uploads/2017/12/NSS-Final-12-18-2017-0905.pdf

White, Karen, "Improving Health During Global Media and Information Literacy Week," Twitter Blog, October 24, 2018.

"Why Is Hate Speech and Fake News Against Okinawa Spreading? Is There Hope for the Internet Era? 'Fact Check' Discussion (2) [なぜ沖縄に対するフェイク情報、ヘイト言説が流れるのか? ネットの時代に希望はあるのか? 「ファクトチェック」座談会【2】]," *Ryūkyū Shimpō*, May 24, 2019.

Wong, Edward, "How China Uses LinkedIn to Recruit Spies Abroad," *New York Times*, August 27, 2019. As of August 28, 2019:
https://www.nytimes.com/2019/08/27/world/asia/china-linkedin-spies.html

Wong, Tessa, "Singapore Fake News Law Polices Chats and Online Platforms, BBC, May 9, 2019. As of July 9, 2019:
https://www.bbc.com/news/world-asia-48196985

Worldometer, "Current World Population," webpage, undated-a. As of March 20, 2019:
http://www.worldometers.info/world-population/

Zhang, Lim Min, "Fighting Fake News Here with Legislation," *Straits Times*, May 13, 2019. As of July 9, 2019:
https://www.straitstimes.com/opinion/fighting-fake-news-here-with-legislation

Zhilin, Gennadiy, "Information-Psychological Weapons: Yesterday and Today [Информационно-Психологическое Оружие: Вчера и Сегодня]," *Soldier of the Fatherland [Солдат Отечества]*, No. 57, 2004.

Zushin, Yevgeniy Georgievich, "Power Has No Equal in Strength [Власть, не имеющая равных по силе воздействия]," *Independent Military Review [Независимое военное обозрение]*, No. 16, April 30, 1999.